THE ACOUSTIC SENSE OF ANIMALS

THE ACOUSTIC SENSE

OF ANIMALS

WILLIAM C. STEBBINS

Harvard University Press
Cambridge, Massachusetts
and London, England   1983

LIBRARY OF CONGRESS CATALOGING IN PUBLICATION DATA

Stebbins, William C., 1929–
The acoustic sense of animals.

Bibliography: p.
Includes index.
1.   Hearing—Physiological aspects.   1.   Title.
QP461.S68   1983   591.1′825   82-21350
ISBN 0-674-00326-8

To Katie

ACKNOWLEDGMENTS

I AM GRATEFUL to Mike Beecher and Art Popper for their suggestions as well as for their support. Janice Jones and Sue Pierson deserve special credit in the typing and careful editing of the manuscript. I particularly appreciated the encouragement and editorial counsel from William Patrick and Maria Kawecki. The Sensory Physiology and Perception Program of the National Science Foundation supported my research at the University of Michigan. It was this research—thinking about it, designing it, doing it, and most of all trying to convey my own excitement about it to my students—that led to this book. To those students who have always been such first-rate colleagues I owe special thanks.

CONTENTS

THE ACOUSTIC SENSE OF ANIMALS

1

SOUND AND HEARING

SOUND is pervasive and ubiquitous. Unlike light it can penetrate solid, opaque barriers, and more easily than light it moves around such objects with little loss in energy to stimulate the comparatively simple and resolute mechanoreceptors of the auditory system. Then, too, sound can act over a considerable range. Very different from other forms of stimulation, it can impart information on current events at an unseen distance—events that may be critical in the life of the listener because they signal an approaching predator, an unsuspecting prey, or a member of the same species whose life is in danger. It is little wonder that there are no habitats which preclude the sense of hearing, or vertebrate species that can do without it, and it plays a key role in the life of the most active and successful of the invertebrates—the insects.

This book is about hearing, which is a behavioral response to the form of energy that we call sound. In looking at how animals hear and what they listen to, we are exploring man's heritage with regard to hearing—the probable course of evolution of a sense which originally was adaptive because it enabled its possessor to respond to events outside its immediate environment and take appropriate action. It often meant the difference between obtaining a meal and becoming one. The sounds were those of movement, usually by another animal, and were transmitted as pressure waves through the surrounding medium: substrate (that is, the earth), water, or air. Perhaps at the same time, or somewhat later, animals developed a way of generating their own sound, and sound then took on the added function of communication within a species, between members and between kin. It is likely that early in evolution, such "intraspecific" signaling centered on reproductive availability. Later, as forms of social organization became more complex, communication served many social functions and evolved to its most intricate stage: human language.

Our sense of hearing may appear to lack the drama and significance of our visual sense. We often take our hearing for granted because it is always with us; rarely can we, or do we, block it out effectively. By comparison, the consequences of entering a dark theater, turning out the lights at night, or simply experiencing the deepening twilight are spectacular. The appearance of the world around us alters profoundly. Objects which are sharply defined, with high contrast and vivid color, become blurred images with fuzzy edges and in varying shades of gray. We are continually reminded of the urgency of our sight but not of our hearing, which never forsakes us even at night, although on occasion we wish it would. We often find noises intrusive: a barking dog, fighting cats, a wailing siren, and so on. Even with ear defenders or plugs we can only slightly attenuate the noise, but not rid ourselves of it completely. Those who have normal hearing are little able to empathize with the hearing impaired, for their world is completely foreign.

Despite its lack of prominence in our lives, our ever-present sense of hearing has obvious survival value. As no other sense can, it provides a constantly vigilant and always alert early warning system. If there were ever any vertebrates who could function without hearing, they are no longer in evidence to tell their story. Certainly the warning capability of the acoustic sense may be one of its most primitive and fundamental functions, upon which others were later built. If hearing enables us to detect and evade predators, it also alerts those of us with carnivorous appetites to the presence of prey—that, too, a very old and primitive function. Third, hearing serves importantly for many animals in the preliminary stages of reproduction, by helping to locate a mate. In fact, for the mosquito this may be its principal function. Fourth, for animals that are organized into social groups, intraspecific communication sounds play a major role in maintaining group cohesion. Such sounds often grade into one another and differ only in subtle ways, placing increasing demands on hearing and on the auditory system for the discrimination of very fine acoustic detail.

The basic mechanisms of the auditory system detect and receive sound energy from the environment and transduce it into electrochemical energy suitable for transmission by the nerve fibers, which relay it to the brain and subsequently to the effector or motor system, where it is translated into some form of behavior. Auditory receptors are part of a mechanical system designed to yield or give way to sound pressure waves and, in so doing, further prepare the signal for its ultimate reception and transmission within the central nervous system. In invertebrates this feat is accomplished in a single stage; in vertebrates the processing is more complex. It is the function of these receptors to detect changes in sound pressure, which is then either converted directly into nerve energy (invertebrates) or is modified in order that it may be more easily transmitted within the body tissues for transduction into nerve energy at a more central location (vertebrates).

A few basic structural plans for acoustic receptors are observed throughout the animal kingdom. Drumlike membranes, stretched taut, are found in the outer ear of most land vertebrates and in some insects, such as locusts and certain moths. Sensitive hairs, or *cilia,* are seen in mosquitoes and in vertebrate inner ears. Balloon-type structures function to conduct sound to the inner ear in some teleost (bony) fishes, and as acoustic receptors in certain moths. Although these structures vary in exact form and function in different species, they are all effective in the detection and transmission of sound waves, particularly in air and water. Whether detection of vibrations in solid or substrate actually constitutes hearing is in some contention. The peripheral structures engaged in the detection of such stimulation (the proprioceptive organs located under the ant's knee, for example) are ordinarily quite different from those considered as typical auditory receptors and will not be considered further here.

The book's title implies a comparative approach to the study of animal hearing; yet comparison for its own sake is of little value, and of interest, perhaps, only for those animals whose ears rank high on some dimension relative to all others. Comparison is important because it may tell us something about the evolution of hearing in animals, about the relevant selective pressures from the environment that have played a role in shaping hearing function, and about the evolution of that closely related feature, communication, which in one species became language.

We cannot, strictly speaking, study the evolution of hearing simply by studying living animals, all of which are, of course, the successful end products of long evolutionary histories. It makes little sense to suggest that the hearing of humans has descended from that of modern fishes; however, by using available fossil evidence on earlier forms together with embryological data, comparative studies on modern animals can afford valuable insights into the evolution of hearing.

ACOUSTICS AND THE NATURE OF SOUND

Sound itself can be thought of quite simply as a mechanical disturbance in air, liquid, or solid by some body set in motion or vibration.
Although the ways by which sound is measured and its course in certain environments can become highly complicated and technical, for
our purposes sound can be considered in fairly clear and straightfor-
ward terms. Simple sounds can be characterized as very small, continuous, predictable, and often repetitive changes in atmospheric
pressure. The simplest of sounds is a pure tone, a single acoustic frequency (whence we derive its pitch); it represents a fluctuating
change in pressure and may be portrayed as a wave undulating
through time and space, as shown in Figure 1.1. Its peaks represent
compression and its troughs rarefaction (dispersion) of molecules.
Like pendulums, the molecules move to and fro when they are in an
elastic medium: air, water, or solid. The effect is passed on from as-
semblages of molecules to their neighbors as a baton is passed from
runner to runner in a relay race. The molecules themselves are not
traveling from sound source to listener, but their effect is, and sound
is in this way effectively propagated through space. Energy is thus
transmitted by a compressional or longitudinal wave of pressure outward in all directions from the source in the form of an expanding
sphere, distributing energy over an ever-increasing area. Molecules

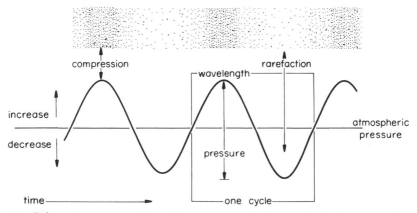

1.1   Schematic representation of pure tone, indicating periodic changes in its
sound pressure with time. Molecules (shown at the top of the figure) are
compressed at maximum amplitude (pressure) of the wave and are separated
(rarefaction) at minimum amplitude. The wavelength—one complete cycle—
is measured between corresponding points on the waveform.

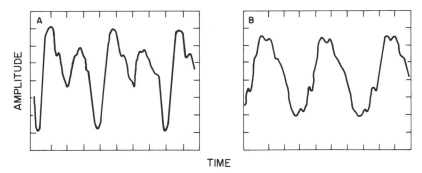

TIME

1.2 Changes in sound pressure (amplitude) over time for a saxophone (A) and clarinet (B). Both instruments are playing middle C—about 260 Hz. (After Burns 1973.)

close to the source disturb neighboring molecules at a greater distance; these subsequently interact with still more distant molecules; and so on. At any point on this expanding wavefront the distributed energy is diminishing in proportion to $1/r^2$, where r is the radius of the sphere or the distance of that point from the sound source. The relation is known as the inverse square law.

Complex sounds, if periodic and repetitive, are no more than complex versions built on the simple wave model for a pure tone described above. These complex sounds consist of many frequencies, all varying in time and bearing a predictable relationship to one another. Bird song, a musical instrument, a baby monkey's call to his mother—all reveal the complexity of sound. The unique qualities of each sound enable the listener to identify the source and behave accordingly. Sounds, then, are combinations of frequencies and intensities: that is, some frequencies (often lower ones) are enhanced, while others may be attenuated. We easily distinguish among musical instruments because of the different emphases at different frequencies. A saxophone and clarinet sound very different, although they both may be playing the same fundamental note. Their relative sound patterns, as they might be visualized by electronic means on the cathode ray tube of an oscilloscope, are seen in Figure 1.2.

Other common biologically relevant or meaningful sounds may have a mushy or strident character. They are not periodic and cannot be broken down completely into simple tonal components. In nature many of these are transient, such as the snap of a twig or the bark of a dog. Nevertheless, these sounds often contain frequencies or bands of frequencies which are intense relative to others. A small dog emits a high-pitched bark, whereas a large dog gives forth with a "deep" growl. We perceive loudness changes not only in the level of the over-all sound, but also in certain frequency regions. A scream, for exam-ple, can become more intense and at the same time appear more shrill (higher frequencies emphasized). The important properties of all sounds are their *intensity* (or, to be more precise, their pressure), their *frequency* (which we often interpret as pitch), their *duration,* and their *timbre or complexity,* which refers to the amount of fre-quency or spectral information they contain. Sensitive instruments are available that can measure these various characteristics precisely.

Sound pressure may be measured in a variety of units: microbars, dynes/cm^2, Newtons/m^2, or Pascals. To further complicate our lives, a special logarithmic scale using a unit called the decibel (dB) is most often used. The number of decibels tells us by how much a certain sound exceeds a given reference level measured in microbars or one of the other units of pressure. The reason given for the use of the decibel scale is that our hearing encompasses such an enormous range of in-tensities that a simple arithmetic scale would produce unworkably large numbers (six or seven digits). The decibel scale compresses the arithmetic scale by a logarithmic transformation and gives us num-bers that seldom exceed two digits. In studying the hearing of most animals, the reference levels 20 μN/m^2 (20 micro Newtons per square meter) or 0.0002 dynes/cm^2 (2×10^{-4} dynes per square centimeter) are most often used. These are equivalent reference levels. In the study of fish hearing, however, the reference levels traditionally em-ployed are 1 dyne/cm^2 or 1 microbar, which are likewise equivalent. These higher sound pressure reference levels for fish reflect the fact that fish are generally somewhat less sensitive to sound than other vertebrates. These different reference levels will appear in the figures throughout the book.

The use of decibels and their reference levels may be clarified by the following example. The normal human speaking voice commonly ranges between 40 and 60 dB SPL (sound pressure level), or, what is the same, 40–60 dB above the reference level of 0.0002 dynes/cm^2. At 40 dB SPL the sound pressure would be 100 times the reference, or 0.02 dynes/cm^2; at 60 dB SPL it would be 1,000 times the reference, or 0.2 dynes/cm^2. Many normal animal sounds are within that range, and some are considerably quieter. Most intense sounds, at least those which exceed 100 dB SPL, such as jet or engine noise or amplified music, are usually man-made and need not concern us here. The deci-bel intensity scale takes some getting used to; a dB scale for common sounds is provided in Table 1.1 to give the reader a frame of reference.

Table 1.1. Decibel scale for common sounds.

Sound level (dB SPL)	Source
0	Audible threshold for primates at 1,000 Hz, wing movement of an owl
10	Rustling leaves
20	Soft whisper
30	Purring cat
40	Monkey "coo" call
50	Bird singing nearby
60	Wolf howling in the distance
70	Barking dog
80	—
90	Roaring lion
100	—
110	Thunder

The table must be considered only a guide, since the distance between listener and sound source is a major determinant of the sound pressure level of the signal at the listener's ear.

The frequency of a sound, or the number of vibrations made each second by the sound source or vibrating body, is measured in cps (cycles per second) or its more modern equivalent Hz (Hertz). Duration is measured in units of time. In evaluating the timbre or complexity of a sound, every attempt is made to describe as much as possible about the spectral and intensive information it provides.

All of these measures are brought to bear in the following instance. Infant and young juvenile macaque monkeys use in their wanderings an expressive signal, the "coo" call, to maintain contact with kin that may be momentarily out of sight. The call, which is often answered and thus reestablishes contact, is tonal in nature but moderately complex. Its *fundamental frequency* is the lowest frequency emphasized and is what gives the call its characteristic pitch. However, there are other frequencies in the call which are often attenuated relative to the fundamental. These frequencies are harmonically related to the fundamental (that is, they are arithmetic multiples of the fundamental frequency). An example of a contact or coo call is pictured in the form of a *sonograph* (commonly known as a voice print) and presented in Figure 1.3. Its overall sound pressure level is about 40 dB above the common reference (0.0002 dynes/cm^2).

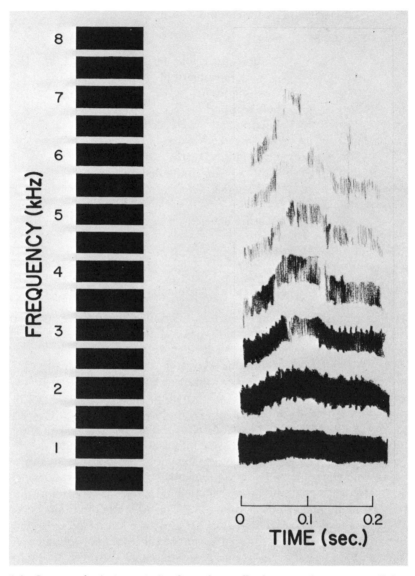

1.3 Sonograph (voice print) of monkey call, showing changes in call frequency (in kHz, or kilohertz) with time. The dark bands represent concentrations of acoustic energy at the different harmonics, which are multiples of the lowest band, the fundamental. (After Green 1975.)

The graph itself simply plots the changes in the frequency of the call over time. The fundamental frequency is centered at about 1kHz (1,000 Hz—middle C is about 260 Hz) at the start and is overlaid with several harmonics. A rise and fall in frequency in all of the bands in the figure are seen before the halfway point in the call, which has a total duration of about half a second. The darkness of the trace is proportional to the sound pressure level of the call but does not provide a quantitative measure of call intensity. Instruments such as spectrum analyzers are available for strict quantification of the properties of an acoustic signal. The sonograph offers a somewhat more qualitative picture of the signal and makes quick, simple, visual comparisons of different signals feasible; it is widely used in the analysis of biological acoustic signals. In the above example we have dealt with the intensity or sound pressure of a signal, its basic or fundamental frequency, its duration, and its timbre or complexity in terms of its various frequency components: the harmonics, their relative intensity levels, and the change in frequency or frequency modulation (FM) occurring during the signal. In examining the hearing of animals, we will be attending for the most part to the variables of signal frequency and intensity, for they are basic to an understanding of hearing, and it is these about which we know most. Only recently have scientists begun to look at the effects of more complex signals on the hearing of animals, because they are so very difficult to treat and to measure either in the context of field studies or in the laboratory. Some of the more recent work in this challenging field will be discussed later.

In discovering how animals hear, how hearing as a sense has evolved, and what adaptive strategies various animals have employed, there are additional properties of sound that should be considered. Although these will be taken up appropriately in the context in which they occur, a few of the most important are mentioned here. The propagation of sound varies according to the medium. In water, for example, the velocity of sound is about 1,500 meters per second, or nearly five times that in air (340 meters per second). In solid substances— bone, wood, and earth or substrate—sound travels even faster. Current conditions such as temperature, humidity, turbulence, water depth, and so on are important determining variables. Vegetation can have significant consequences for the propagation of sound. Transmission characteristics will vary greatly from meadow to forest floor to forest canopy. Further, in a complex sound the low frequencies suffer relatively less attenuation than the highs. Thus, a sound that at its source consists of a wide range of frequencies may be reduced considerably in intensity and in frequency bandwidth at the ear of the distant listener. The implications of all these factors for the hearing and communication of animals is tremendous, and yet this is a field of scientific endeavor that is still in its infancy. In evolving unique adaptations to their environment, do animals take advantage of some of these known characteristics of sound? For example, do animals appear to concentrate their vocal energies at low frequencies for long-distance signaling? Is low-frequency hearing characteristic of species that must hear acoustic events at a considerable distance?

As sound travels through a medium, it encounters objects which are of a very different composition from that of the medium of the original sound source. In traveling through air, sound is, to an extent, reflected and absorbed when it contacts trees, walls, animals, and so on. Reflected sound, as in an echo, is an indispensable tool for some animals for purposes of navigation and orientation. Bats in the night air, and porpoises and other sea mammals under water, emit signals which they subsequently receive in the form of reflected sound. The accuracy with which these animals are able to resolve the smallest imaginable targets (bats, for example, can perceive tiny moths) rivals the visual acuity of many vertebrate species.

All structures have a fundamental frequency at which they will vibrate, once set in motion by a transmitted pressure wave in the medium or by direct coupling to a vibrating body. If the frequency of vibration of the disturbances from the medium or vibrating body is close to the natural or fundamental frequency of the structure, the structure will *resonate*—that is, vibrate at its natural frequency. Because of the properties of the structure, the amplitude of vibration is enhanced and, in addition, may occur at multiple integers—harmonics—of its natural frequency. Thus, a structure whose fundamental frequency is 1 kHz may exhibit modes of vibration, usually of diminished amplitude, at 2, 3, and 4 kHz. The result is a complex sound resembling the call shown in Figure 1.3. Of interest to bioacousticians is the fact that resonance effects occur in enclosed or partially enclosed volumes of air and are a function of the size and shape of the enclosure. The swim bladder, which performs the function of conducting sound to the inner ear of certain fish, and the larynx and external ear canal of mammals are examples of biological enclosures which use the resonance effect to advantage in hearing and communication.

THE INVESTIGATION OF HEARING

Hearing is what we make of sound; in effect, we define hearing as the behavioral response of an animal stimulated by sound. Such a definition is concrete, precise, and operational, and it enables us to observe and examine the acoustic sense of animals by dealing directly with their behavior in response to sound. Such behavior may be observed under natural conditions: the moth altering its flight path in response to the echo-ranging signals of the bat, the shark homing in on the sound of an object thrashing in the water, or the Old World monkey answering the contact or clear call of its close relative, for the moment out of sight. We also examine an animal's hearing in the laboratory under more artificial but much more carefully controlled conditions. We condition an animal to respond to sound by reinforcing it with food for doing so. We can then test its auditory acuity or the limits of its acoustic resolution by presenting a wide range of stimuli varying in intensity and frequency and measuring the animal's ability to respond over this range. These are psychophysical experiments, similar to those carried out with human subjects. The response by the animal, which often entails pressing a small button at the onset of acoustic stimulation, is designed to be equivalent to the human linguistic responses "Yes, I hear it," "A sounds louder than B," and so on. Such laboratory experiments, as well as the observations made of animals under natural conditions, will be described in the chapters that follow. In both instances it is the measurable and observable responses to sound which provide the data—the objective evidence for the wide-ranging and diverse auditory abilities of animals.

Among the variety of ways in which animals use their hearing, perhaps the most widely understood is their capability for detecting minimal sound levels—their *absolute threshold*. Behavioral procedures in the laboratory enable us to determine auditory thresholds over a range of frequencies to which animals can respond. Such measures reflect an animal's sensitivity to sound and how that sensitivity may vary with sound frequency. In the case of some animals, where behavioral methods are not practical, small wires (electrodes) may be inserted into the inner ear or auditory nerve in order to record the small electrical voltage changes which occur when the ear is stimulated by sound. These measures are less than valid as quantitative measures of hearing, but they do provide an estimate of an animal's audible range and sensitivity, and, for some animals, that may be all we can obtain at the present time. Behavioral measures of threshold or electrophysiological recording from stimulated inner ear or auditory nerve provide us with much of what we know to date about the auditory capabilities of animals. Armed with these data, we can begin to gain some knowledge of the form and function of the acoustic sense. For example, the fascinating but bewildering array of forms or anatomical structures which subserve hearing in animals assumes a more dynamic biological reality when we know something about the hearing that goes with them. Transmission and scanning electron microscopes have begun to reveal these structures in their most intimate detail.

We are led to inquire about the function of the acoustic sense in the animal's own world. How is an animal's sense of hearing parlayed into successful strategies which have ensured that animal's survival and reproductive success to the present day? Exciting discoveries in field biology have shown us the continuing acoustic duel between the predatory, insectivorous bat and its prey, the night moth, or the desert rat's successful use of auditory cues in avoiding predators from air or ground.

There is more to know about an animal's hearing than sensitivity and frequency range. For example, the fact that an animal is sensitive acoustically and has an extended audible range fails to inform us about its discriminative capabilities within that range. Somewhat complicated procedures have been employed on birds, mammals, and even some fishes to determine their ability to discriminate between closely adjacent sound frequencies or intensities. These are abilities which we take more or less for granted, for we use them daily in linguistic encounters and certainly in the enjoyment of music. Another aspect of hearing which is essential to survival is the ability to locate accurately the source of a sound. Sound localization has been observed in animals in the field and actually measured in the laboratory. Its complexity has made the procedures difficult, and only a few animals have been studied thus far.

Acoustic discrimination, sound localization, and other important attributes of hearing can give a more complete picture of an animal's acoustic capabilities. Although many are known and have been examined in humans and a few other mammals, their wider application awaits a patient investigator and further improvements in techniques of study.

Three explicit features of the acoustic sense of animals will occupy our attention in this book. Behavioral and physiological measures of hearing will be supplemented by a description of the relevant morphology of outer and inner ear structure. The use to which hearing is put in the lives of animals will be the third and most directly relevant feature of our concern. It is hoped that the reader will gain not only an appreciation of what and how animals hear and why they listen, but also an understanding of the nature of scientific inquiry and the rewards to be gained in unraveling significant and interesting biological problems.

2

INSECTS: THE AUDIBLE

INVERTEBRATES

THE DIVISION of the animal kingdom into invertebrates and vertebrates is not the way the more rigorous taxonomists would have it; in fact, for our purposes, it may even be considered an otocentric or "ear-centered" view of taxonomy. There are few, if any, vertebrates who lack at least a rudimentary sense of hearing. On the other hand, although most invertebrates are able to respond to some form of displacement or vibration of the medium in which they live, we would be stretching things a bit to refer to this as hearing. Of all the many phyla of invertebrates, it is the arthropods, particularly the insects, who make the most use of sound and who, in fact, depend upon it in their daily cycle of eluding those who would eat them, eating others, and reproducing with their own kind. In some insects, sound may be used in territorial defense; in insects that swarm, sound may promote cohesion.

Insects hear with ears which bear little outward resemblance to those of vertebrates. These ears are in no way homologous receptors implying common ancestry, for the insects and land vertebrates have evolved independently for hundreds of millions of years; yet they are analogous because they share the same auditory function. Only a limited number of designs have evolved by which live tissue—skin (or insect cuticle), cartilage, and bone—has adapted for the effective reception of sound and its subsequent transduction to nerve energy. For example, a thin region of insect exocuticle stretched taut serves as a particularly sensitive form of acoustic receptor for the locust, and in principle reminds us of the tympanum, or ear drum, so common in the land vertebrates.

We study the ears of insects because by so doing we can gain a better comprehension of the role of hearing in all animals. It is a salient component of an animal's behavior, as are seeing, eating, fighting, mating, and moving. General principles regarding acoustic form and function emerge, contributing significantly to the understanding of our own sense of hearing.

In addition, the hearing of insects forms an exciting story in its own right. Those with a more practical turn of mind will be interested to know that in the future it may be possible to control insect pests acoustically, given an adequate knowledge of their auditory capabilities and their uses of sound.

ACOUSTIC RECEPTORS AND THEIR FUNCTION IN INSECTS

Among the simplest insect hearing organs are the *hair sensillae*—filamentous hairs, often long, which are lightly hinged at their point of attachment to the integument, or insect cuticle. These sensitive, freely moving hairs respond to wind direction, but also to low-frequency sound stimulation. Their function resembles that of a mechanical lever; that is, the hairs are long relative to their area of attachment to the cuticle, so that a small force widely distributed over the length of the hair is translated into a much larger force at the small region of attachment. A single hair may be hinged in such a way as to permit movement in only one direction; other hairs hinged differently provide potential information regarding the direction and location of a sound source.

Although these hair sensillae are seen in many insects, it is unlikely that they all mediate responses to sound at reasonable biological levels. There is convincing evidence that they do, at least in certain species of caterpillar, grasshopper, and cockroach. Hair sensillae are pictured in Figure 2.1 on the anal cerci (rear end) of a cockroach, where their most likely function is an early warning system against predation from the rear.

Early experimenters, in attempting to verify the role of hair sensillae in hearing, would load the hairs with flour or water droplets, or even smear the animal's body with Vaseline. In most instances, behavioral or physiological responses to sound disappeared or were attenuated but quickly returned when the material was removed. Frequencies to which these receptors appear most sensitive are well below 1,000 Hz.

2.1   Sensillae on the rear end of a cockroach; these hairs are sensitive to low-frequency sound waves. (After Autrum 1942.)

More complex and intricate hairlike receptors are found in the form of *Johnston's organ* in the mosquito and the feathery *arista* of the fruit fly. Both receptors are an integral part of the antennae located on the animal's head. An artist's rendering of Johnston's organ, clearly shown to have an auditory function in the male mosquito, can be seen in Figure 2.2. This highly specialized receptor structure extending from the head is located between the second and third antennal segments. *Fibrillae* (hairs) extending from the flagellar shaft are set in motion by sound, causing the entire shaft to vibrate. The lever analogy seems equally appropriate in accounting for the effective mechanical operation of Johnston's organ and the variety of hair sensillae. As with the hair sensillae, evidence for the importance of the fibrillae or flagellar shaft in the detection of acoustic stimuli was demonstrated by their removal or by preventing their action by fixation or loading with beads of shellac. Responsiveness returned when the shellac was removed.

Mosquitoes have been the subject of relatively intense investigation, due in part to their capability for carrying yellow fever as well as their general nuisance value. In the species that have been examined most carefully, it is the low frequencies (200–400 Hz) produced by the wing beat of the adult female which serve to attract the male. Male mosquitoes will orient toward and approach flying females and perform the initial part of the mating response, consisting of seizing and clasping. It has been shown in the laboratory that they will respond in the same way to an artificial sound source set at the frequency of the flight tone of the female; that is, they will orient toward the source and make the seizing and clasping response. The admirable parsimony of nature is reflected in the fact that it is simply the frequency of the female flight tone that the male responds to at distances of up to 25 centimeters. Given two sources of identical frequency, the male will choose the more intense.

Thus, it appears that the antennal receptor of the male mosquito,

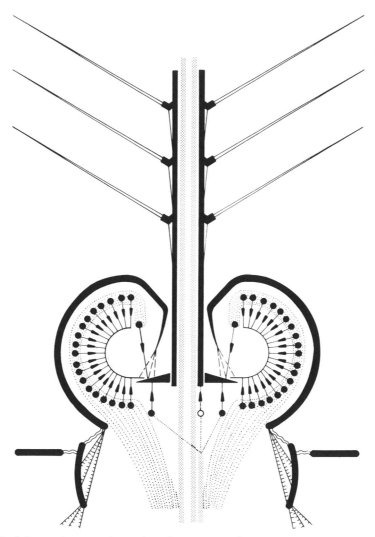

2.2   Johnston's organ, located on the antenna of a mosquito; it is sensitive to low-frequency sound as produced by the wing beat of the female of the species. Three hairs, or fibrillae, are shown extending from each side of the flagellar shaft in the upper part of the figure. In the lower-middle section of the figure is the pedicel, a spherical structure, split in half in order to show the sensory cells, which are represented by black circles, and nerve fibers, represented by dotted lines. (After Risler 1955.)

2.3   The feathery arista, which serves as the acoustic sense organ in the fruit fly. The arista receptor, with its many branches, is seen against the backdrop of the fly's eye. (After Burnet et al. 1971.)

among the most sensitive insect acoustic receptors of its kind, functions only in the specific behaviors which are precursive and critical in mating and reproduction. The flight tone of the female is an attractant in the initial stage of courtship; other sensory systems probably take over in the later stages of mating.

Sound serves a similar function in some species of *Drosophila,* the fruit fly, although the sex roles of male and female are reversed relative to the mosquito: the flight tone of the male fly attracts the female. The male stands on a flat surface about 5 millimeters from the female, faces her obliquely, extends one wing, and vibrates it. The feathery arista of the female is sensitive to the wing tone of the male, which at 5 millimeters has a sound pressure level of only about 50 dB, equivalent to the loudness of conversational voices at a distance of several feet. The male is at an advantage, for he can express his intent to a particular female without fear of being overheard by other flies in the vicinity. As in the mosquito, this acoustic biosignal is specific and critical in the initiation of reproduction. An arista receptor, part of the fly's third antennal segment, is shown in Figure 2.3; like the mosquito's, it is a delicate and sensitive detector of nearby sound. It is also a structure of undeniable beauty.

Probably the most highly developed of the insect acoustic receptors, and in many ways the most interesting and best understood, are the tympanal organs found in the crickets, locusts, katydids, cicadas, moths, and butterflies. These detectors range widely in complexity from the relatively simple but highly specialized organ of the night moth, with only two receptor cells, to that of the cicada, with more than one thousand. Tympanal receptors are paired organs consisting of a *cuticular membrane*—a very thin region of the exoskeleton located in the abdomen, over the posterior chest cavity, under the wings, or below the knee. The membrane typically covers a *trachea,* or air cavity; hence the derivation of the word *tympanum,* or drum. On the body, it is often located in a deep cleft which has sometimes been compared to the outer ear canal in mammals. Such an ear is seen in the night moth in Figure 2.4.

Sensory cells are usually directly attached to the membrane's inner aspect, as in the night moth (Figure 2.5). These cells, together with their supporting tissue, were appropriately named *chordotonal organs* by a nineteenth-century biologist, Vitus Graber, because many of them stretched like taut strings or cords from point to point. Graber, among others, believed that they served an auditory function. Although many of them are auditory receptors in some insects, in others they may serve as chemically sensitive receptors (*chemoreceptors*) or even as receptors which provide information to the animal regarding the position of its body or limbs (*proprioceptors*). For that reason it is often unwise to generalize about the function of a particular receptor structure in insects. Furthermore, in one insect a single structure may contain more than one kind of receptor.

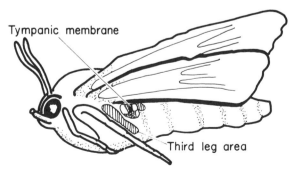

2.4   Night moth, with position of right ear canal indicated under the wing. (After Ghiradella 1971.)

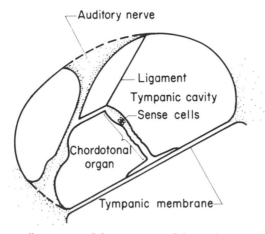

Auditory nerve

Ligament
Tympanic cavity
Sense cells

Chordotonal organ

Tympanic membrane

2.5 Schematic illustration of the inner ear of the night moth. The position of the tympanic membrane and the attachment of the acoustic sensory cells and auditory nerve fibers are shown. (After Roeder 1967.)

If the biomechanics of the sensilla-type receptors resemble the action of a lever, the tympanum, with its attached chordotonal organ, acts more like a piston. In response to sound pressure changes the tympanic membrane moves, driving the smaller-diameter chordotonal organ as a shaft (see Figure 2.5). The force distributed over the relatively large expanse of membrane is amplified to a greater force acting on the smaller area of the attached chordotonal organ. The amplification factor or mechanical advantage is related, then, to the ratio of the size of the entire tympanic membrane to the restricted area of attachment of the chordotonal organ.

The basic structural design of the insect tympanal organ reminds one of the sound-conducting pathway in the middle ear of the land vertebrates. As we shall see later, the resemblance to a piston, the areal ratios, and the mechanical coupling between a large membrane and a small shaft are evident in the tympanic membrane and columellar shaft of amphibians, birds, and reptiles and in the ear drum and bony (ossicular) chain of mammals. In insects the sense cells are affixed to the tympanic membrane, and the conversion of sound into the electrochemical energy of nerve fibers is accomplished in one step. In some way, still poorly understood, the biomechanical action is sufficient to initiate activity in the sensory afferent nerve fibers, which then transmit the information to higher centers in the nervous system for subsequent action. In vertebrates the process is somewhat more complicated. Whereas the tympanic membrane is located either on the body surface or close to it, the sense cells are contained deep within the inner ear and their activation depends on a sequence of energy transformations which start at the membrane.

MOTH VERSUS BAT

Each of the bilateral tympanal organs of the night moth are linked to only two sensory cells. The elegant work of Kenneth Roeder and his colleagues over the years has revealed the fascinating biological relationship which has evolved between prey and predator: the moth and the insectivorous bat. The relationship is based on the moth's ability to evade the bat with the aid of acoustic cues and only two sensory cells on each side. Laboratory recordings of the electrical activity of these two cells have shown that they are most sensitive to ultrasound between 25 and 60 kHz, which encompasses the auditory frequency region used by neighboring bats while hunting with their echolocating cries. Further, the two cells are functionally different: one is more sensitive than the other by 20–30 dB, as indicated by their electrical activity in response to sound.

Field observations, together with the results of the laboratory experiments, strongly suggest the following scenario. The more sensitive cell of each pair provides an efficient early warning system, while the less sensitive cell is still beyond the range of the echolocating bat's ultrasonic sonar, at a distance of about 30–40 meters. The more sensitive cell, in an apparently reflexive manner, signals the flight muscles of the moth's wing. The rapidly executed and well integrated behavioral response is an appropriate change in flight direction which increases the distance of the moth from the bat and, in so doing, reduces the signaling activity of these more sensitive cells.

The important question of directionality, or localization, was next addressed by Roeder and colleagues. Given that the moth could register the approximate distance of the bat with time to maneuver, what sort of information was available regarding the direction of the bat's approach to enable the moth to take appropriate evasive action? Clearly, as in all animals with two ears and with some distance between them, the sound will arrive at the ear nearer the sound source slightly earlier, and will be slightly more intense than at the far ear. Further, since the sound arrives later at the far ear, it will be in a different part of its cycle (phase), thus creating a third potential cue for localizing its source. The relationships are complex and depend on the characteristics of the signal as well as those of the head of the listener, particularly on the distance between the ears.

When electrical activity to ultrasound stimulation was again recorded from the more sensitive receptor cells on each side, the moth's
wing position was found to be an important variable. With wings
raised above the horizontal plane, the near ear (relative to the sound
source) was found to be 30–40 dB more sensitive than the far ear. The
greatest difference between the two ears occurred when the sound
source was at right angles to the long axis of the moth's body. Consider that the moth's ear lies under the anterior portion of its wing. As
the wings drop below the horizontal plane, the side-to-side (lateral)
intensity difference favoring the near ear gives way to a top-to-bottom
(dorsoventral) intensity difference, such that sounds from above are
10–25 dB less intense than those from below. With a wing beat cycle
of 25 milliseconds, the moth is provided with an almost constant
source of information about the direction in three dimensions of an
ultrasound source, based essentially on these two different disparities
in sound intensity. Less obvious, perhaps, is the moth's ability to obtain directional information from a source directly to the front or to the
rear. An increase or decrease in the activity of the sense cells with distance from the sound source is a plausible answer.

The physiological data which provide such compelling evidence for
directional hearing in the night moth were obtained using the ingenious experimental arrangement shown in Figure 2.6. The moth was
carefully positioned at the top of a small tower in an anechoic chamber
(a room constructed to be both soundproof and free of echoes). While
the experimental measurements were being taken, the tower could be
rotated in the horizontal plane and the overhead speaker in the vertical plane. Thus, the data were obtained over the surface of a complete
sphere with the moth at the center and the speaker at a fixed radius.
The measurements were repeated for different wing positions.

On those occasions when the moth is picked up on the hunting
bat's sonar at, say, a distance of only 2–3 meters, the chase is on. The
less sensitive receptor cells now change their firing rate and signal the
moth's flight muscles. The subsequent behavioral response, very different from that elicited by the more sensitive, "long-range detector"
cells, may take a variety of forms, all of which appear erratic, even un-
predictable, to the observer and perhaps also to the bat. In most instances the behavior is directed toward the ground, which probably
increases confusion in the returning echo signal of the bat. Small objects such as leaves, twigs, and berries are easily confounded with the
moth. Often the moth will dive directly to the ground with wings mo-
tionless. Occasionally the trip downward will include a wide variety of
flight maneuvers like those of a stunt plane: loops, rolls, and one or
more banks or light turns. As the story goes, Roeder himself first no-

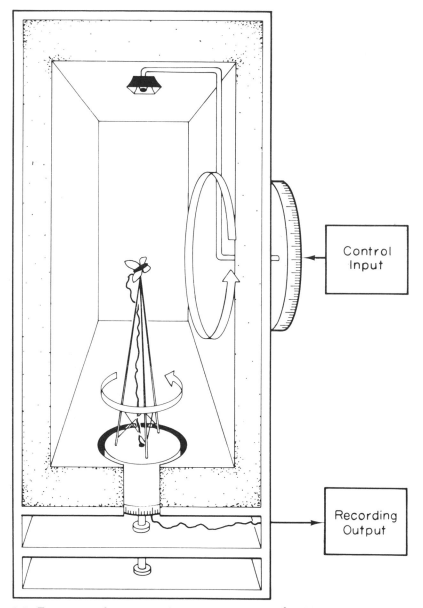

2.6   Experimental apparatus for measuring directional hearing in the night moth. (After Payne et al. 1966.)

ticed this behavior during an outdoor party on a summer evening when one of the guests was jingling some keys, which are capable of producing high-frequency acoustic stimulation.

Other moths show strategies very similar to those of the night moth. It seems reasonably certain, at least in these species, that the evolution of the relatively simple but highly specialized tympanal organ is based on its sole function as a detector of the acoustic signals emitted by the moth's chief predator, the echolocating insectivorous bat.

Evasion, on the other hand, may not be the only defensive strategy used by moths against bats. For example, the arctiid moths exhibit flight maneuvers very similar to those of the noctuids, or night moths, when stimulated by real or synthetic bat echo signals. But arctiids possess an effective second line of defense. When acoustically stimulated by ultrasound, including bat cries, these moths emit an extended series of ultrasonic clicks which appear to prevent attacks by bats. Whether these sounds act to jam the bat's sonar or are intense enough to be repellent to the bat is not yet clear, but the effectiveness of these clicks in turning away bats is not in question.

FREQUENCY DISCRIMINATION

We have seen convincing evidence that insects can hear, in the sense that they are able to detect a wide range of sounds in their environment. It is also clear that certain moths, for example, are capable of localizing the source of a sound in space, and even of distinguishing between two levels of sound intensity. Although this may seem primitive relative to the uncanny hearing of many of the higher vertebrates, insects have to be given special marks in view of their small body size, their marginal nervous system, and the acoustic equipment they have to work with. In view of these considerations, it is not surprising that although the hearing of many insects covers an extensive frequency range, there is little evidence that insects are able to discriminate between the frequencies within that range. Recent anatomical and physiological studies by Axel Michelsen have shown that the locust is indeed capable of such discrimination, and that it probably can divide its audible frequency range into three, and perhaps four parts. How this ability serves the locust in its everyday commerce is much less clear, although Michelsen and others have offered some suggestions.

The locust (*Acrididae*) has a tympanal organ on each side of the first abdominal segment. Its basic structure reflects the role it plays in

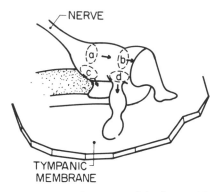

2.7 The Müller organ, an acoustic sensor of the locust. Four separate groups of receptor cells (a, b, c, d) are attached to the inner surface of the kidney-shaped tympanic membrane. (After Michelsen 1971.)

frequency discrimination. Attached to the tympanic membrane's inner surface at different locations are four separate and distinct groups of receptor cells, shown in Figure 2.7. The so-called *Müller organ*, like other tympanic organs, transfers a relatively small force, distributed over the entire surface area of the membrane, to the smaller region of attachment of the sensory cells.

Using laser holography and an electronic capacitance probe, Michelsen studied the characteristics of the moving tympanic membrane (2.5 × 1.5 mm at its widest part) as it responds to sound stimulation. According to Michelsen, the amplitude of vibration of the membrane differs at different locations on its surface, and, further, the differential pattern of movement changes with the frequency of the acoustic stimulus. As a consequence, the four groups of receptor cells attached at different positions on the membrane encounter differing modes of vibration, thus providing a physical basis for frequency discrimination in the locust ear.

Finally, with the aid of a very fine glass microelectrode, Michelsen was able to record electrical activity to varying frequencies of sound in those four groups of cells. Although there is some overlap, the different groups are evidently maximally sensitive to different frequency bands of stimulation, a fact that provides firm evidence for their role in frequency or pitch discrimination in the locust ear. Additional evidence, also based on electrophysiological recording, indicates that much of this peripheral frequency information from the receptors is retained in more central portions of the locust nervous system. The

use to which the locust puts its ability to discriminate sound frequencies is uncertain, although it is likely to be related to the task
of discriminating the songs of conspecifics from those of other
species.

EVOLUTIONARY CONSIDERATIONS

In the course of evolution, the insects, among the invertebrates, have
yielded to certain selective pressures by developing a variety of super-
ficial receptors, sensitive to acoustic events and able to cope with at
least the initial stage of transduction in the auditory process. The reproductive advantage gained in so doing has been substantial. The degree of success in the detection and interpretation of acoustic signals
ranges from modest or marginal to substantial, as in the example of
the sophisticated locust tympanal organs.

There are doubtless other forms, yet undiscovered, which have
evolved as acoustic receptor organs. Many will probably resemble in
detail those described here. Often, the progressive evolution of such
structures from organs serving a proprioceptive function to ones that
can detect simple substrate movement or changing air currents, and
eventually to ones that can perceive aerial sound, appears feasible on
the basis of the evidence from comparative anatomy. The discovery of
new and very different forms underscores the variety of structural designs capable of meeting the minimum standards as acoustic detectors.

All known acoustic organs, from the relatively primitive but nonetheless effective systems of insects to the highly evolved mammalian
middle and inner ear, function basically as mechanoreceptors, structurally designed to detect acoustic signals for transduction and then
transmission within the nervous system. Flexible hairs or thin membranes exposed to the external environment are free to move with
changes in sound pressure—that is, as a consequence of molecular
displacement in the surrounding medium. In simpler systems, such as
those of insects, transformation of these energy changes into propagated bioelectric potentials in the nervous system may be the next
step in the process; by contrast, in the mammalian ear an elaborate
chain of events intervenes between movement of the tympanum and
the final stage of transduction at the synapse of the sensory hair cells
with the auditory nerve fibers within the *cochlea*—a spirally coiled
bony tube in the inner ear.

As with any mechanical system, there is no simple one-to-one correspondence between what goes in and what comes out. Energy losses occur due to friction and to changes in the medium through which the sound waves are propagated (for example, air to fluid-filled inner ear in terrestrial mammals). What is significant is the manner in which different auditory systems have evolved. Selective pressures have led to the development of acoustic transducers which, with the constraints imposed by the characteristics of the organism to which they belong, are designed in ingenious ways to compensate for these frictional and other losses. Amplitude, wavelength, timbre or complexity, temporal pattern, and phase are properties of sound waves to which animals respond. The successful evolution of acoustic receptors is predicated upon their fidelity in receiving and transmitting these properties of acoustic signals from the environment.

What evidence there is suggests that the selective pressures from the environment which fostered the development of a variety of acoustic receptors in the insects were often highly specific. In many species, receptors evolved to serve very limited but explicitly defined functions. On the other hand, the receptors have become sufficiently specialized so that they carry out these functions very well indeed. Such a degree of specialization is exemplified in the ear of the night moth, which apparently functions solely as a bat detector, and in the Johnston's organ of the male mosquito, sensitive to the frequencies in the wing beat of the female which are an attractant leading to the first stage of reproductive activity.

Arthropod acoustic receptors, unlike those of vertebrates, are primary sense cells; that is, they are true nerve cells, an extension of the nervous system, and therefore they combine the multiple functions of detection, transduction, and transmission to more central structures in the nervous system. Because the skeletal design of insects imposes limitations on body size, the insect nervous system appears to have sacrificed the complexity necessary for abundant information and decision making in favor of speed. The economy and time saving that result from primary sense cells, large fibers, few synapses, and relatively direct sensori-motor connections are evidence that these adaptations are well suited for rapid, evasive action.

The auditory systems of vertebrates clearly did not evolve from anything resembling the invertebrate sense organs we have described. Any similarities that we do find, such as tympanic membranes, are examples of convergent evolution. The evolution of the vertebrate ear from the invertebrate ear is shrouded in mystery. In fact, there are few speculations as to what might have occurred so many millions of years ago. One possibility, suggested on more than one occasion, is that the

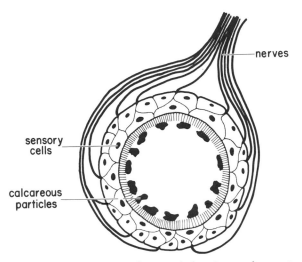

2.8 The statocyst organ, a gravity detector belonging to the marine mollusk *Helix*. The sensory cells with overlying calcareous particles are sensitive to the animal's positional changes relative to gravitational forces. (After Wells 1968.)

vertebrate inner ear was the eventual outgrowth of a nonauditory organ in invertebrates, a device that provided sensory feedback to invertebrates concerning their position in space. The gravity sensor, which permitted these animals to maintain equilibrium and which is seen in modern coelenterates and other marine invertebrates, is called a *statocyst*. Although it takes different forms in various animals, its basic plan is the same as that in the mollusk *Helix*, illustrated in Figure 2.8. The statocyst is a fluid-filled sac suspended in a fluid-filled cavity and surrounded by cartilage close to the brain. In the inner sac are dense, relatively heavy, calcareous particles that are attached to the ends of cilia. These hairs are an integral part of cells which are innervated by nerve fibers. As the animal turns over or simply moves with respect to gravity, the heavy particles exert a force in a new direction on the hair cells, thus providing the animal with information concerning head position and motion.

The statocyst hypothesis suggests that the more sophisticated vertebrate labyrinth—the vestibular portion (our organ of balance or equilibrium) containing semicircular canals, and sensory tissue with hair cells as receptors in a fluid-filled medium close to the brain—derived from something very similar to the invertebrate statocyst. It is then a relatively short step from the vertebrate vestibular organ, the upper or superior part of the labyrinth, to its more recently evolved lower or inferior extension, the cochlea. The statocyst hypothesis is an intriguing one and not without its logical support, but it is very unlikely that we will ever accrue the data necessary to alter its status as "only" a hypothesis.

3

THE ENIGMATIC FISHES

BEFORE WE CONSIDER hearing among the fishes, it is necessary to break this large and diversified group into more meaningful and useful taxonomic entities, for fishes make up four of the eight major subdivisions or classes of vertebrates. First there are the primitive jawless fishes, the agnathans, represented by the living lampreys and hagfish. Second are the armor-plated, jawed fishes of the Paleozoic, the placoderms, which are extinct. Third are the cartilaginous fishes, or chondrichthyes, which include the sharks, skates, and rays. Fourth are the bony fishes, the osteichthyes, which make up the largest number of living fishes, including the modern teleosts of whom a subgroup, the ostariophysines, have been identified on the basis of their unique and very effective adaptation for hearing. There is no evidence of hearing in either the agnathans or the placoderms, but it is quite possible that the agnathans, at least the lampreys, are able to hear, for they do possess anatomical structures similar to those that subserve hearing in other fishes. There is little doubt that cartilaginous fishes respond to sound, and many bony fishes hear very well—some particularly acutely.

AUDITORY STRUCTURES IN FISHES

Hearing in fishes has presented an unusual enigma and a long-standing controversy and disagreement which is only partially resolved even now. How sharks perceive sound, which they seem able to do, is still very much a mystery, for example. There are at least two aspects to the problem. First, until recently, the behavioral techniques for measuring hearing in fishes were not available. Second, many fishes possess two distinct receptor systems, both sensitive, though not equally so, to underwater sound. To understand how fishes hear and to appreciate the controversy surrounding their ability to do so, it is important to grasp the basics of these two receptor systems.

Fishermen have long thought that fishes are possessed of the most sensitive hearing, and that the best fisherman is a quiet fisherman. In order to hear human speech at normal conversational levels in air, fishes would have to be more acoustically sensitive than we are, since less than 1 percent of the energy of aerial sound is transmitted from air into water. The interface between any two media such as air and water provides an effective barrier to sound. While some sound is transmitted across this interface, a considerable amount of sound energy is also reflected. However, footsteps in the bottom of a boat are transmitted to fishes more efficiently, much as is the disturbance produced by a stone or stick thrown into the water.

Along the body of most fishes, and in the region of the head, run long stripes or lines visible to the eye (see Figure 3.1). These lines represent canals which at intervals are open to the water. Protected under cover of, along, and athwart these canals are viscous, gelatinous, tent-shaped structures (*cupulae*) containing hairs embedded and rooted in sensory cells (*neuromasts*), which are mechanoreceptors sensitive to changes in rate of water flow along the canals. Figure 3.1 shows a magnified cross section of a lateral line canal with the neuromast in the path of the flow of water along the canal and attached to nerve fibers beneath. The cupula, along with the hairs, is sensitive to the direction of water flow, and the information is transmitted via the sensory nerves to the brain. These sensory structures are capable of detecting minute water displacements. They not only provide the fish with feedback on the direction of its own swimming movements, but also register any changes in water turbulence that might be due to friend or foe, prey or predator. There is considerable controversy over whether the neuromasts are sensitive enough to pick up low-frequency sound from a nearby source—sound that results from water displacement created close to the source and that dissipates rapidly with distance. The effect is quite different from the sound pressure wave, which is transmitted effectively over far greater distances and is the mode of stimulation to which the ears of most animals are sensitive.

The second receptor system is the labyrinth, or inner ear, which in

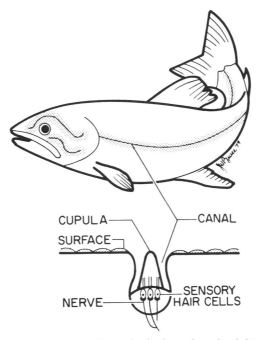

CUPULA———
SURFACE—
————CANAL

NERVE—
———SENSORY
HAIR CELLS

3.1 Lateral line canals shown along the body and in the fish's head region. Inset below presents a magnified cross section of one canal, illustrating the position and structure of the acoustically sensitive hair cells and nerve fibers embedded in the gelatinous cupula.

vertebrates is a euphemism for those structures inside the head that are associated with balance or equilibrium together with those that function in hearing. Figure 3.2 is a schematic drawing of the labyrinth of a bony fish. The upper, or superior, part includes the *semicircular canals* and *utricle*, structures that play a major role in fish equilibrium. Interestingly, the sensory structures in the completely enclosed and fluid-filled canals closely resemble the neuromasts of the lateral line system in both structure and function. The striking similarity has led many to suggest that the evolution of the vertebrate labyrinth was due to a gradual enfolding and departure from the body surface of a more primitive lateral line canal system. The lower, or inferior, part of the labyrinth includes *saccule* and *lagena*—organs that in many bony fishes contain hair cells as receptors for hearing. In the land vertebrates (tetrapods), a large extension—called the *cochlear duct*—of the lower saccule takes over the function of hearing, while the utricle, saccule, and semicircular canals all become primarily balance organs.

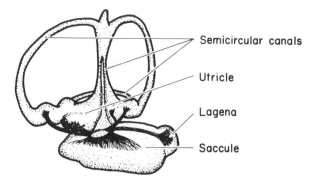

Semicircular canals

Utricle

Lagena

Saccule

3.2   The labyrinth of a living fish, showing the position of the organs of balance, the utricle and semicircular canals, together with the hearing organs, the saccule and lagena. (After Retzius 1881.)

The mechanisms and pathways by which the inner ear (not including the organs of balance) is stimulated vary greatly among different groups of fishes; in some they are still unknown. Modes of stimulation will be considered below as different fishes and their hearing are discussed.

Sharks. Behavioral evidence based on field observations indicates that sharks are sensitive to acoustic stimulation, for they can home in on underwater low-frequency sound from a distance of several hundred meters. Because of the distance, visual cues are ruled out, and olfactory (smell) stimulation is unlikely. However, it is also unlikely that the neuromasts of the lateral line organs, no matter how sensitive, could detect the extremely minute water displacements that are generated at such considerable distances. The existence of an inner ear system in sharks is strongly suggested by the behavioral findings, although the structures responsible have not been determined unequivocally. A likely candidate for a peripheral sensor is an area of depression, the *parietal fossa,* located on the top of the shark's head and covered with tightly stretched skin, resembling a drumlike membrane. Experiments have shown that a small region of sensory cells, the *macula neglecta,* located near the saccule, is especially sensitive to vibratory stimuli applied directly to the parietal fossa. Whether these structures form part of a "true" auditory pathway for underwater acoustic stimulation is yet unknown.

The necessary behavioral experiments are difficult to conduct, not merely because of the inherent risk to the scientists but also because of the problems in ruling out stimuli other than those the experimenters are manipulating. The tests are usually carried out on the open ocean in areas known for their heavy shark populations. Acoustic underwater speakers are lowered from a small boat into the water. First, the number of sharks is observed with speakers silent. A few are attracted, probably by the activity of the boats and the lowering and positioning of the speakers. Gradually the number of sharks in the area wanes; then low-frequency signals are transmitted over the speaker, and the number of sharks in the immediate area increases dramatically. In fact, for a few minutes after the speaker has been turned off, new sharks continue to appear in the proximity of the speaker. Such accurate homing, even after the signal is no longer present, raises some interesting questions about the additional sensory and even neural capabilities of these animals. The observational data gathered by two caged human observers in the water beside two vessels 200 meters apart strongly suggest that sharks respond to acoustic signals at that distance by approaching the sound source, although it is difficult to know with complete certainty if the sharks sighted leaving one observer's area are the same as those sighted approaching that of the second observer. The signals which attracted the sharks in greatest number were in a frequency range of 25–50 Hz and intermittent (pulsed at a high rate) rather than continuous. There is a strong similarity between this form of acoustic energy and an animate object thrashing in the water. Experiments such as these have been carried out with great care many times and by many different scientists, in both the Atlantic and Pacific. If, as seems most likely, the sharks are responding to acoustic stimuli received by an auditory sytem other than the lateral line, several very interesting questions await future exploration. How do sharks sense these acoustic events, and what structures—peripheral and internal or central—mediate this reception? Second, what are the mechanisms which permit the shark to pinpoint the location of the sound source so unerringly? And, third, how is this ability retained for as much as several minutes after the acoustic signal has been turned off?

Bony fishes. There is ample evidence that many of the bony fishes are able to hear; some have hearing which almost rivals our own at certain sound frequencies. Although fishermen have always known this to be true, it was not until fairly recently that scientists could agree that bony fishes, which constitute most of the fresh and salt water fish familiar to us, possessed a sense of hearing as opposed to a mere sensitivity to water displacement through the lateral line organs. There are several reasons why it semed improbable to many nineteenth- and early-twentieth-century biologists that fishes could hear.

First, most fishes appeared mute, and it seemed unlikely that a sense of hearing would exist in a quiet, noncommunicating animal. We know now that some fishes frequently produce audible sounds. But because the appropriate underwater sound measuring equipment had not yet been invented, and because much of the energy in underwater sound is not transmitted across the water-air interface, this information was not available to early investigators.

Second, the inner ear of mammals has two distinct structures, each with a unique function. The semicircular canals, the utricle, and the saccule are organs of balance; attached to the inferior part of the saccule is the large snail shell, or cochlea, which was well known as the organ of hearing in mammals (see Chapter 5). Since fishes had no cochlea, they were considered incapable of hearing. As we know now, the saccule in many bony fishes (the utricle in some) and neighboring lagena function very differently in fishes from the way they do in mammals, and contain the sensory cells that respond to sound and transmit acoustic information to the brain.

Third, many scientists were unable in the laboratory to demonstrate behavioral evidence of hearing by fishes. To those who could provide such evidence, those who could not argued either that the results were due to artifact (a lack of control in the experiment), or that the behavior observed was not a reaction by the fishes to sound pressure changes but to the movement or displacement of water. Certainly, many early experiments lacked proper controls, and fishes that "heard" could have been responding instead to other types of stimulation.

The biologist Karl von Frisch pointed out that the sound sources used in early attempts to demonstrate fish hearing—a violin, a singer's voice, a harmonica, a whistle—had little biological relevance in the world of fishes, and that therefore we should not expect the fishes to respond to those sounds. Von Frisch proceeded to whistle to a catfish, at the same time dropping food in its tank. The fish, which was blinded so that it could not respond to visual cues, soon became conditioned to the whistle and responded to it by immediately swimming to the surface of the water. The experiment represents one of the earliest attempts in the laboratory to use a conditioning procedure to examine a fish's hearing.

In the course of evolution, it is likely that in the inner ear of bony fishes (Figure 3.2) the function of hearing supplanted the more primitive role of gravity detection. A typical receptor organ located in the saccule or lagena is shown in Figure 3.3. This macular organ is made up of sensory cells (mechanoreceptors) and bears an interesting re-

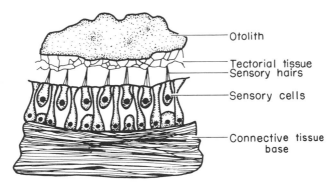

Otolith

Tectorial tissue
Sensory hairs

Sensory cells

Connective tissue
base

3.3 A typical receptor organ in a living fish. Note the relatively heavy ear stone, or otolith, atop the delicate hairs of the sensory cells, which are embedded in a base of connective tissue. (After Wever 1974.)

semblance to the gravity detector shown in Figure 2.8. It is noteworthy that similar structures serve the different functions of hearing and of maintaining equilibrium, and it is likely that the basic anatomy of the receptor has not changed significantly since its incorporation into an acoustic mode. Some increase in its sensitivity is likely, along with a change in its innervation pattern and connections within the nervous system.

Sound waves under water enter the fish's body with little resistance, for it is approximately the density of water and therefore transparent to sound. The sensory cells, supported on a base of connective tissue, are set in motion in response to the incoming sound waves. The *otolith* (ear stone), which is heavy and of greater density than either water or the tissues of the fish, is also set in motion but with less amplitude and less abruptly, and is therefore out of phase with the movements of the sensory cells. Because the sensory hairs are connected to the otolith by the tectorial tissue, there is a bending of the hairs—a mechanical deformation—due to the differential motion between the sensory cells and the otolith. It is widely believed that this bending is sufficient to excite the sensory cells to action, so that they may transmit the necessary information to the central nervous system along the nerve fibers they serve.

It is likely that the hearing of fishes would never have amounted to much without a more efficient sound-conducting pathway to the inner ear. The differential stimulation due to the otolith's being heavier than the surrounding tissue does not provide a very sensitive acoustic receptor. Fortunately, in the course of evolution, structures which had probably evolved as adaptations to nonauditory systems were later exploited by the auditory system, much to its advantage. One of the most important of these was the gas-filled swim bladder, generally located in the abdominal cavity between spine and intestine and behind the region of the head and inner ear (see Figure 3.4). Although its primary or more primitive role may have been as a monitor of hydrostatic pressure, it may also have served in buoyancy, respiration, and sound production—a remarkable variety of important functions.

The swim bladder, almost like a sensitive balloon, expands and contracts in response to sound pressure waves which enter the tissues of the fish from the surrounding water. The pressure wave is thus converted to one of displacement, which is then transmitted to the nearby inner ear containing saccular and lagenar otoliths. Although the swim bladder represents a considerable advance over the relatively crude otolith system alone, some fishes have evolved still further improvements in the conducting pathway between swim bladder and inner ear. In the herring, for example, the swim bladder has long tubular extensions which terminate in the immediate proximity of the ear. There is, consequently, less energy lost in the intervening tissue between swim bladder and ear, and the sound conducting route is thus more efficient.

Among the bony fishes the ostariophysines have developed a particularly effective sound pathway (see Figure 3.5). The three or four most anterior vertebrae in the spine have been modified and provide a

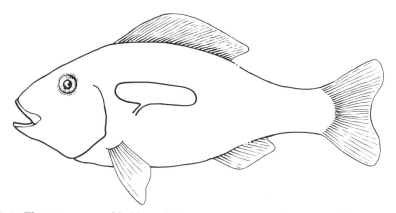

3.4 The swim or air bladder, which receives and conducts sound in certain species of living fish, is shown in its position close to the head and inner ear.

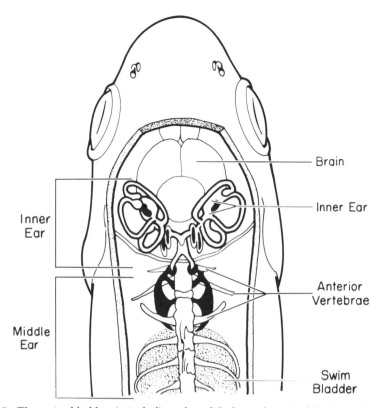

3.5   The swim bladder (stippled) and modified vertebrae (in black) in the ostariophysines act as a conducting link, enabling sound to be transmitted from
the surrounding water to the inner ear. (After von Frisch 1936.)

direct link or bony (ossicular) chain between the anterior swim bladder and the auditory part of the inner ear. The movements of the expanding and contracting bladder are transmitted to the ossicular
chain and hence directly to the canals of the inner ear, or auditory labyrinth. It is among the ostariophysines (the catfish, goldfish, and carp,
among others) that we find some of the most acoustically sensitive
living fishes.

It was the early-nineteenth-century investigator Ernst Weber who,
with consummate dissecting skill and the eyes of an eagle, first attrib-
uted the function of this modified vertebral ossicular chain (which
now bears his name) to hearing. It was not until well into the next
century, when behavioral measurement techniques were developed,
that it was possible to give further evidence of the function not only of

the ossicles but of the swim bladder, saccule, and lagena as the receptive centers for sound in the fish inner ear. Von Frisch performed a series of delicate operations on the auditory system of the minnow, an ostariophysine, for the purpose of removing or damaging certain parts in order to determine their normal function. His experiments and those of others helped confirm the logical and well reasoned speculations of the early-nineteenth-century scientists.

Additional evidence for the function of the saccule and lagena in fish hearing has been provided by recording the electrical activity evoked by sound stimulation in these receptor organs and in the nerves which innervate them. In the typical experimental arrangement, electrodes (extremely small wires, or glass micropipettes filled with potassium chloride) are placed in contact with the appropriate structures: saccule, lagena, or nerve. The fish, which is anesthetized and restrained in a special holder, is stimulated by sound, and the ensuing electrical activity that is picked up by the electrodes can then be recorded.

HEARING IN FISHES

Perhaps the most convincing proof of fish hearing comes from the behavioral experiments, many of them conducted in the laboratory but some in the open water. Experiments performed little more than two decades ago left little doubt that fishes were capable of hearing. However, early attempts were often flawed by incomplete or incorrect measurements of sound under water, a problem that was exacerbated by the use of small tanks which often produced complex effects on the acoustic waveform. Ambient noise levels—that is, environmental noise present during the experiments—was often not considered, and may have masked or virtually obliterated the acoustic stimuli presented by the experimenter. In addition, as mentioned above, only recently have techniques been available for conditioning fishes to respond to sound, so that we might probe the limits of their hearing acuity with the necessary precision and reliability.

If fishes can hear, what is their auditory threshold—the least amount of sound energy to which they can respond reliably at different sound frequencies? And what range of frequencies does their hearing encompass? A single function answers both of these questions. In Figure 3.6 the threshold function for a marine teleost (modern bony fish) obtained over a range of pure tone frequencies is presented, together with the human audibility function as a standard of comparison. The fish in question is a squirrelfish of the genus

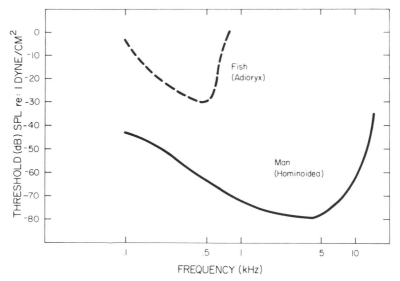

3.6   Auditory thresholds taken over the range of audible sound frequencies for a modern bony fish, compared to those of man. (After Coombs and Popper 1979; Sivian and White 1933.)

Adioryx, which possesses a swim bladder but no obvious conducting link between swim bladder and inner ear. Its threshold function, or audiogram, is almost V-shaped, with greater sensitivity slightly below 500 Hz and a useful hearing range extending from about 100 to 800 Hz. We may gain some insight into this animal's auditory acuity relative to our own, without the usual anthropocentric (human-centered) reflections on man's superiority.

To determine such sensitivity functions in man, it is a matter of instructing the subject in a standardized manner and accepting as data simple "Yes" ("I hear it") or "no" answers. Many determinations are made in which the acoustic stimulus is varied in sound pressure level and frequency. The resultant threshold functions represent stable averages, and the data points on the function are referred to as the minimum detectable levels of sound pressure at each frequency to which the subject is able to respond. In animals, a behavioral conditioning procedure is necessary in order to prepare them to respond to sound in the test environment reliably and accurately.

One such conditioning procedure uses a brief, mild electric shock which is preceded by an acoustic stimulus. This stimulus acts as a warning of impending shock, which can be avoided if the fish responds to it by crossing a barrier that bisects the fish tank. Thresholds are obtained by lowering and raising the stimulus intensity many times and determining the intensity value to which the animal responds correctly 50 percent of the time (the traditional definition of threshold or minimum detectable sound level). In the skilled hands of someone with knowledge of behavioral techniques, the procedure is workable and the data reliable. Occasionally, the shock can produce emotional behavior which may result in highly variable threshold estimates or, at times, a subject that becomes immobilized and will not respond.

A second procedure also uses shock, but the conditioning paradigm is quite different. The fish is not required to respond, in the sense of taking a certain action as in the avoidance procedure. Shock by itself causes a brief but marked slowing of the breathing rate. If an acoustic stimulus directly precedes shock, it also in time can briefly slow respiration. The process by which the acoustic stimulus acquires this function is known as classical or Pavlovian conditioning. Once conditioning is established, threshold testing may proceed in the same manner as used in the avoidance procedure.

A third procedure, discussed in greater detail in Chapter 5, entails a food reward which is contingent upon responding correctly to auditory stimulation. With mammals, food reinforcement has proven effective without the debilitating emotional behavior so often engendered by electric shock. It has not enjoyed a similar success with fishes, partly due to the difficulty in maintaining motivation over a lengthy test session, and partly due to food delivery problems underwater.

Members of the same squirrelfish family as *Adioryx* are species of the genus *Myripristis,* which possess a conducting link from swim bladder to inner ear in the form of two long tubular extensions of the swim bladder. Their hearing threshold function, shown in Figure 3.7, may be compared with that of *Adioryx.* The differences in sensitivity (about 20 dB) and frequency range are striking, and it is a reasonable assumption that the advantage to *Myripristis* is related to its more efficient conducting pathway. Its hearing extends from at least 100 to 3,000 Hz, with greatest sensitivity in the range of 400–1,500 Hz. Among those fishes which have been studied, only certain of the ostariophysines with a bony conducting chain between swim bladder and inner ear appear to have more acute hearing, with a frequency range which extends, at least in the catfish, beyond 6,000 Hz. By mammalian standards this would be considered fairly primitive hearing, but underwater it serves these animals very well and permits a simple intraspecific auditory communication system to function effectively.

Although sensitivity and frequency ranges set the limits of detecta-

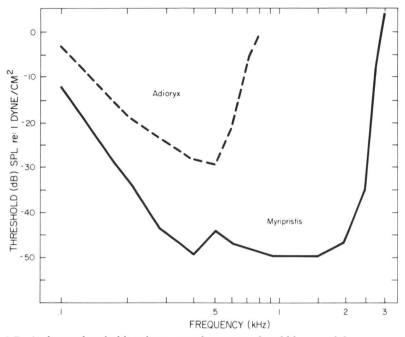

3.7 Auditory thresholds taken over the range of audible sound frequencies compared for two bony fishes. The more sensitive *Myripristis* (solid line), has an effective conducting link between swim bladder and inner ear which *Adioryx* (broken line) lacks. (After Coombs and Popper 1979.)

bility on an animal's hearing, they tell us very little about its capacity for discrimination within these limits. Discrimination ability has important implications for the reception and decoding of sounds more complex than the stimuli employed in threshold testing. These are sounds which the fish must attend to in its everyday commerce with its environment and upon which its very survival depends. With the obvious distance-sensing limitations on underwater vision and smell, the evolutionary pressure not only to hear at some distance but to discriminate one sound source from another must be intense.

One measure of discriminative acuity entails conditioning the animal to respond to changes in the acoustic frequency of the stimulus. The frequency shift acts as a warning signal for electric shock, and the fish learns to avoid on this basis or is conditioned such that its respiration slows markedly during the frequency change. Frequency discrimination thresholds are then determined by narrowing the gap between frequencies until the subject is able to respond correctly to the frequency difference half the time.

Ostariophysines such as the minnow and the goldfish, with sound-conducting bony ossicles from swim bladder to inner ear, are able to detect a change in frequency of about 20 Hz from a 500 Hz tone; a smaller shift in frequency is not discriminable. Considering the simplicity of the structure of the fish ear relative to that of mammals, the acuity is quite remarkable. Differential frequency thresholds for these fishes lie between those for human skin (actual vibratory frequency difference thresholds measured on skin) and those for human hearing. In the few nonostariophysine species that have been studied, differential acuity is very poor relative to the ostariophysines. Although the connection between swim bladder and inner ear may be important, differences in the inner ear and the innervation pattern of the auditory nerve also may contribute significantly to the fine frequency-resolving power of the ostariophysine auditory system. Until more behavioral and morphological data are obtained from different species, these conclusions must remain tenuous.

One of the most important functions of hearing for a fish, or for that matter for any animal, is its ability to locate and identify the source of a sound and act accordingly. Prey, predator, potential mate, conspecific rival—all must be treated differently, and the accuracy of identification depends both on an animal's sensitivity to sound and its acuity for discriminating the various dimensions of acoustic stimulation (frequency, intensity, and so on). However, the actual localization of the source of a sound (pinpointing an object in three-dimensional space) is a different matter and must somehow depend on differential stimulation of two or more structures in the ear. In insects with paired tympanal organs or in tetrapods or land vertebrates with two distinct and separate *pinnae* (ear flaps), outer ears, and middle ears, sounds from a given direction will reach the ear nearest the source slightly before reaching the far ear. Certain sounds incident at one ear will be out of phase with the stimulation at the other ear because of the time lag produced by the distance between the two ears. Further, the head itself acts as a sound shadow at higher frequencies, such that the sound reaching the far ear is less intense than that reaching the near ear. Thus, in land vertebrates the cues provided by differences in time of arrival at the two ears, phase, and intensity permit the accurate location of a sound source (see Chapters 4 and 5).

The mechanism for localization is not readily apparent in fishes, for they lack the two external ears, one on each side of the head, which are common to land vertebrates. Also, since sound travels so much faster in water, the cues based on time of arrival and phase difference are considerably reduced. Further, because of the relatively small

head size in most fishes, the head's function as a sound shadow is unlikely; and the fish's body is close enough to the consistency of water
so that it fails to act as an effective barrier to sound waves. Yet the experimental evidence is unequivocal: some fishes are able to localize
sound with reasonable precision.

The potential for directional sensitivity has been established by
physiological recording of the electrical activity in the inner ear of
fishes in response to sound stimulation from different directions. In
addition, behavioral experiments have shown that some fishes will
swim directly toward, others away from, an underwater speaker that is
producing sounds similar to those the fishes themselves might experience. And conditioning experiments have gone so far as to provide
quantitative data on the acuity of fishes in localizing sound. For example, a fish can be conditioned to respond when a sound stimulus
moves along an arc in either the horizontal or vertical plane. The extent of the angular movement in degrees of arc which the fish is able
to detect provides an objective measure of its localization acuity.

An interesting experimental arrangement for just this purpose is
shown in Figure 3.8. The investigation was conducted off the west
coast of Scotland. The equipment shown here was placed on an underwater tower 6 meters above the seabed and 20 meters below the surface. The fish was placed in a small cage facing the lowest acoustic
speaker. A hydrophone (underwater microphone) monitored the
sound near the fish. The fish was conditioned to respond when the
sound shifted from the lowest speaker to another speaker in the vertical plane. The fishes (in this series of experiments, cod were used)
were able to detect a change of as little as 16 degrees of arc in the vertical plane. In a similiar experiment, their angular resolution in the
horizontal plane was found to be 20 degrees.

Although our own ability is somewhat better than this in the horizontal plane, we find it almost impossible to localize sound in the vertical plane, perhaps a reflection of our more two-dimensional environment in comparison with the fish's three-dimensional one. As we have
suggested, the mechanism for directional hearing in fishes must be
very different from that found in insects or tetrapods. The swim bladder in modern bony fishes is a single structure located in the midline.
Its stimulation by sound and the subsequent transference of that stimulation equally to the two inner ears is unlikely to provide the fish with
any directional information. The lateral line organ, with its canals on
either side of the body and head, is clearly capable of being differen-
tially stimulated according to the direction of a sound source, but
probably only at very low frequencies and effectively at a distance of

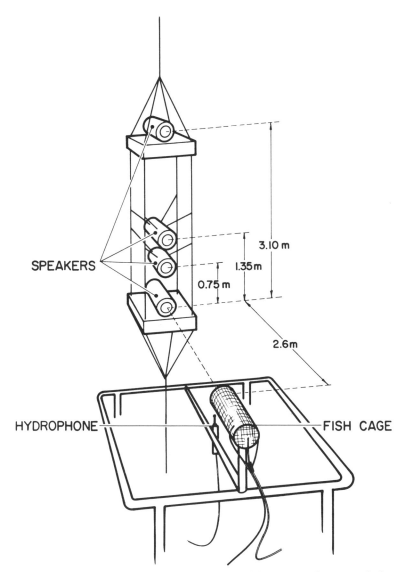

SPEAKERS

3.10 m

1.35 m

0.75 m

2.6m

HYDROPHONE

FISH CAGE

3.8 Apparatus for measuring fishes' directional hearing in the vertical plane, under the ocean surface. (After Hawkins and Sand 1977.)

2–3 meters. The answer may lie in the recent anatomical discoveries concerning the *maculae* (patches of sensory receptor cells) and their differing orientations in the fish, and the orientation of the hairs on the sensory cells themselves. It is entirely possible that the hairs are either more sensitive or exclusively sensitive to the direction of an acoustic disturbance. Thus, their displacement may provide the fish with information regarding the location of an acoustic source. It is not clear that such a directionally sensitive arrangement would function over very great distances. Localization of sound and the morphological and physiological mechanisms by which it is accomplished remain one of the central problems in the study of underwater hearing.

COMMUNICATION IN FISHES

What fishes listen to is still largely an unresolved question. One would be safe in assuming that generally, as with any animal, sounds important to the fish's survival and reproductive success are those which are heard and to which the fish attends. Sounds produced by prey or predators are candidates, yet the fish is perhaps more likely to sense such intrusions as hydrodynamic variations through the lateral line system. There is increasing evidence that some fishes listen to the biological signals produced by their own species. Communication among fishes does not appear to be as widespread or even as elaborate as it is in birds or mammals, but it does seem to serve certain key functions within particular species.

Compared to light or chemical stimuli underwater, sound has certain properties which enable it to be detected quickly and at some considerable distance from its source. Low-frequency stimuli to which fishes are particularly sensitive suffer little attenuation with distance relative to light stimuli and are propagated at higher rates then chemical stimuli. The acoustic signals produced by fishes are usually low-frequency and pulsatile, conforming to fixed or variable temporal patterns. Their nature confirms the importance of temporal cues for the fish as opposed to the spectral or frequency characteristics which are so important to the mammalian auditory system. To us, the sounds produced by fishes may sound like "pops" or "crunches"; they are often generated by the scraping together of bones or teeth, or by the contraction of muscles which rapidly alter the volume of the swim bladder in order to produce audible sound.

One of the most revealing experiments underwater is the taped playback to the fish of its own species sounds: observing the fish's be-

havior enables one to ascribe the proper function to the particular biological signal being played. In other so-called field studies, and even in
laboratory studies, fishes are simply watched in their natural habitat,
and their response to the sounds of conspecifics is documented. Several behaviors in different species have been found to be correlated
with sound emission. Commonly observed in some are what appear to
be territorial and aggressive displays reminiscent of those seen in
many land vertebrates. Sounds produced by males of certain species
seem to inhibit hostile responses in conspecifics, suggesting the possibility of threat displays related to dominance hierarchies. In others, behaviors associated with sound production are related to reproductive
activity. The study of communication among fishes is difficult at best,
and it is not surprising that we know as little as we do about its function in their ecology.

4

AMPHIBIANS, REPTILES,

AND BIRDS

THE SUCCESSFUL TRANSITION of the vertebrates from water to land set the stage for many major, new, and unique adaptations of organ systems and bodily parts to the new medium. Over the millennia the adaptations have continued, and these systems and parts have been further refined to reflect their role in the particular econiche (setting) in which the different vertebrates found themselves. For the sense of hearing the necessary adjustment in auditory structure was considerable, and the major pressure on selection was perhaps related to the differences in the way sound behaves under water, on terra firma, and in the air. The acoustic problem which was to confront those earliest creatures that ventured out of the oceans and onto the land was one of sound transmission from the substrate and from the air to the well protected and insulated inner ear close to the brain. The principal and earliest adaptations that were required were in the sound-conducting pathway or middle ear.

What had served an analogous function as underwater sound detectors in those fishes ancestral to the amphibians would not suffice on land or in the air. Fossil remains reveal little about how those early ancestral fishes solved this problem of sound conduction to the inner ear. Some form of air bladder may have registered pressure waves and transmitted them as displacement to the inner ear much as in modern bony fishes. Perhaps some crude form of sound reception was provided by an otolith (ear stone) in the inner ear through its relative motion with regard to the receptors—the hair cells. By whatever means sound conduction and reception were accomplished by the ancestral fishes, it would have had to be discarded by the amphibians when they left the water. Yet it is possible and even likely that the earliest amphibian structures which served nonauditory functions were, if not cannibalized, at least called into service in the conduction of sound from the ground or from the air to the inner ear, where sensitive membranes overlying delicate hair cells could receive the acoustic information for processing and transmission to the brain and thence to the muscles and glands, where it could be expressed in some form of behavior.

It has been suggested that the early amphibians were able to hear sounds from the substrate by means of bone conduction. Sound conduction paths through the lower jawbone (close to the inner ear) or through the leg and shoulder may have aided these animals in detecting the low-frequency stimulation produced by other large animals in the neighborhood. As these amphibians continued to evolve from crawling to a more erect posture, the sound-conducting pathway likely became too long and tortuous to be efficient, and incoming sounds were masked by the animal's own bodily movements. Proponents of this theory believe that selection pressures favored a more sensitive system—one receptive to aerial sound.

A view that is perhaps more commonly accepted than the one above is that the very early tetrapods already had the rudiments of an ear which would be sensitive to sound transmitted through the air. There is evidence that the fishes immediately ancestral to these amphibians possessed, in the head region, air-filled *diverticuli* (tubes) or pouches which served as auxiliary respiratory organs. On land these air spaces could have been the forerunners of a middle ear cavity. A thin dermal covering near the body surface has been hypothesized, and its analogy with an ear drum is inescapable. Close inspection of the fossil evidence indicates that bones which were part of the rear jaw-support system in many fishes were able to form part of a middle ear sound-conducting pathway in some early amphibians, analogous to the cartilaginous and bony conducting system found in most living tetrapods, including the mammals.

THE AMPHIBIANS

There are three orders of living amphibians: Order Apoda, consisting of wormlike creatures that burrow underground; Order Urodela, made up of newts and salamanders; and Order Anura, including the frogs and toads. Little is known about the hearing of the urodeles or apoda, but based on observations of structure their hearing is thought to be fairly primitive. The apoda ("footless"), although they have an inner ear and some bony conducting elements of a middle ear, possess no middle ear cavity or ear drum. They appear limited to the detection of substrate vibrations at low frequencies. The urodeles also have no middle ear cavity and it is likely that they, too, hear only bone-conducted sound—perhaps from the air, but more likely through the substrate. Neither the apoda nor the urodeles are known to vocalize. The hearing of both groups is probably rudimentary and encompasses little more than sounds in their immediate vicinity.

The adult frogs and toads, on the other hand, are well equipped for aerial hearing; and sound, particularly from their own vocalizations, plays an integral and critical role in their existence. The anurans have been the center of attention of some very interesting experiments on their hearing, and for that reason they will be emphasized below.

Auditory Structures in Amphibians

A significant, even astounding event that takes place in the lives of amphibians is metamorphosis, in which the young larval animals, heretofore living underwater, undergo a dramatic transformation of most bodily parts and are thus equipped for a more terrestrial existence. In the anurans the inner ear remains the same, while the middle ear conducting apparatus is completely rearranged. Near the head region, the tadpole has an air-filled sac (lung) which functions, like the fish's swim bladder, as a sound pressure-to-displacement transformer. During metamorphosis the adult frog or toad acquires a new middle ear (see Figure 4.1) analogous to that found in most living land vertebrates.

In the adult anuran the tympanic membrane or ear drum is a modified patch of skin which lies unprotected at the body surface behind the eye. A bony shaft, the *columella,* is attached at its distal end (the end furthest from the midline of the animal) to the tympanic membrane and proximally (closest to the midline) to the oval window of the fluid-filled inner ear (see Figure 4.1). It is significant that the area of the tympanic membrane is large relative to the area of the oval window, so that the pressure distributed over the larger tympanic membrane is transmitted by the columella to the smaller oval window with a net gain in pressure at the latter. Normally, at an air-fluid interface most of the incident sound is reflected; in this instance the middle ear conducting system acts as an amplifier to restore much of the sound energy that would otherwise be lost in the process of transmission to the fluid of the inner ear.

The anuran inner ear contains two distinct patches (papillae) of

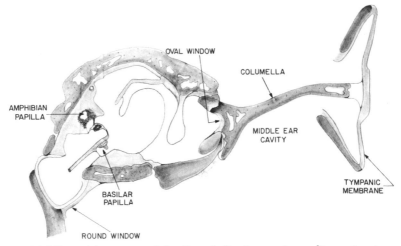

4.1 Middle and inner ear of the frog, indicating major auditory structures. (After Wever 1974.)

tissue sensitive to the displacement of the inner ear fluid when the tympanic membrane–columella–oval window conducting complex is stimulated by sound. The round window (see Figure 4.1) provides a pressure release when the oval window is stimulated. The two papillae, called the basilar and the amphibian, contain hair cells with an overlying membrane (tectorial). Sound-induced movement of the inner ear fluid causes the tectorial membrane to move, and the mechanical energy is transferred to the sensory hairs and thereby to the sensory cells beneath, which rest on a solid base. The peripheral stimulation is complete when, by some process not yet understood, this mechanical form of energy is transduced into the electrochemical energy of the auditory nerve fibers, which transmit the information to the central nervous system and brain.

Hearing in Amphibians

In addition to being separate structures, the two papillae appear to have different functions with regard to their role in hearing. The *basilar papilla,* for example, is sensitive to the higher frequencies audible to frogs and toads (1,200–1,600 Hz). The somewhat larger and more complex *amphibian papilla,* on the other hand, is receptive to lower frequencies in the 200–800 Hz region. More detailed examination has shown that there are two distinct sensitive areas within the amphibian papilla which have important implications for acoustic function. One area contains cells which are responsive to very low frequencies, in the neighborhood of 200 Hz; the other is sensitive to slightly higher frequencies, near 500 Hz. Physiological recording of the electrical activity of individual nerve fibers has confirmed the existence of these three different regions of sensitivity to low, middle, and high frequencies. The frequency limits of these regions will vary somewhat depending on the species of frog or toad. Individual nerve fibers are active in response to low-frequency, mid-frequency, or high-frequency stimulation. Of particular interest is the fact that fibers which are low-frequency (100–200 Hz) sensitive are easily inhibited by the addition of sound energy within the mid-frequency (500–600 Hz) range. Neither the mid- nor high-frequency fibers are easily inhibited by other frequencies or frequency bands.

Communication in Amphibians

The significance of this physiological and anatomical detail in the life of the frog or toad is only recently beginning to be appreciated, and we owe much to the work of Robert Capranica, Lawrence Frishkopf, Glen Wever, and their colleagues for their intensive study of the structure and function of the frog peripheral ear and auditory nervous system. The anuran's ear is closely tuned to its voice, and communication in the anurans plays a major role in their mating behavior. In fact, it is the croaking of the adult male frog or toad which attracts the gravid (egg-bearing) female in the initial part of the mating sequence. The dominant frequencies in the male's croak enable the female to approach and select a member of her own species, and thus the communication signal serves additionally as a species-isolating mechanism. Further, the calls are often geographically distinct, in the same way as are our own regional dialects.

The significant characteristics of anuran calls have been studied in a clever and interesting way in "playback" experiments in the laboratory. A female anuran is placed in the center of a circular arena (see Figure 4.2) with several loudspeakers around the circumference. Any part or all of a natural call which has been recorded on tape, a computer synthesized call, or simply artificially generated combinations of frequencies can be played over a single speaker and the female's approach, avoidance, or indifference noted. In other experiments tape-recorded or synthetic calls can elicit croaking in large captive laboratory colonies.

The acoustic stimuli most effective in producing approach by the

LOUDSPEAKER

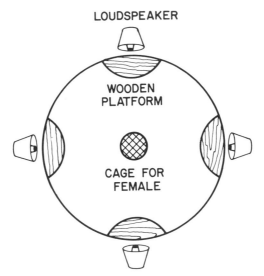

4.2   Arena for playback experiment in frogs. Male call is broadcast from one of four speaker locations. Female is placed in center cage, and her course toward the speaker of her choice is then charted. (After Whitney and Krebs 1975.)

gravid female or vocal responses in the colony have proved to be sounds which contain peak spectral energy (frequencies) in the 100–200 Hz region and in the 1,400–1,600 Hz band. Energy around 500–600 Hz that matches or exceeds the sound energy in the 100–200 Hz band does not generate either approach on the part of the females or croaking in the colony. The physiological recording experiments have shown that it is the frequencies in the 500–600 Hz band which inhibit the low-frequency-sensitive nerve fibers. It has been suggested that this property of the anuran's auditory system may well serve to separate the men from the boys. The young frogs have smaller vocal sacs and mouth cavities and thus generate higher frequencies than the adults. Their calls fail to attract gravid females and to provoke calling in other frogs. As they mature, their vocal cavities grow and their call frequencies correspondingly decrease until they reach the range of effectiveness.

Although the mating call is by far the most common vocal signal observed in anurans, vocal behavior also serves an important function in other contexts. Should an adult male frog be approached by another adult male while the former is engaged in calling for a mate, the calling male will switch to the territorial call. If the intruder does not retreat, the two will fight until one withdraws. Normally, in preparation for mating the male frog clasps the female from behind. If the male instead clasps another male or a nonreceptive female, they will utter "release" calls. In addition, some anurans will emit warning calls when startled and before entering the water in attempting to escape. Although the vocal repertoire of the anurans is limited it serves a critical function in the lives of these vertebrates, particularly with regard to reproduction.

THE REPTILES

The reptiles, in evolving an egg that could be laid on land, carried the transition from water to land one step further than the amphibians. A developmental period in the water was no longer necessary, and the reptiles thus became the first completely terrestrial vertebrates. Although the reptiles are generally thought to have a higher level of organization than the amphibians, there is no simple or direct evolutionary relationship between these two groups. Their common ancestry occurred in antiquity many millions of years ago; thus, both classes (Amphibia and Reptilia) have evolved independently for a very long time. These are important considerations when their hearing is examined, for although gross similarities in structure are evident, reflecting what may have been present in their common ancestry, differences in the fine detail of their auditory anatomy and in their hearing are readily apparent.

Modern reptiles claim membership in one of four orders: Squamata, including the lizards, snakes, and amphisbaenians; Rhynchocephalia, consisting of only one species, the tuatara; Chelonia, made up of the turtles; and Crocodilia, which includes the alligators and crocodiles. Wever, in his very thorough work *The Reptile Ear,* has studied the peripheral auditory structure and physiology of many species of reptiles. Although they all possess some degree of hearing, the variance in auditory sensitivity among them is considerable. Interestingly, those animals which are the most vocal, such as the crocodiles and some of the lizards, seem to have the most highly developed auditory system and probably are the most acoustically sensitive. In turn, this somewhat tentative finding suggests that at least in some of these reptiles there are fairly persistent selection pressures for the development of an effective intraspecific communication system. Unfortunately, we probably know less about hearing and communication in the reptiles than in any other vertebrate group. The work of Wever and of others has set the stage with detailed descriptions of auditory

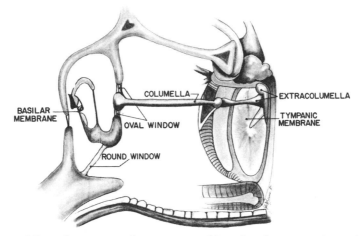

4.3   Middle and inner ear of one species of lizard, indicating major auditory structures. (After Wever 1965.)

morphology and physiology, but our understanding of reptiles' actual sense of hearing and the functional significance of the acoustic signals which they transmit is modest.

Auditory Structures in Reptiles

In the reptile ear, we see for the first time a clear resemblance to the functional mechanical system of the type found in the ears of birds, humans, and other mammals. The middle and inner ear of a lizard are shown in Figure 4.3. Gross similarities between this ear and the anuran ear are obvious. The differences are significant but difficult to discern except at a more microscopic level. As in the anuran, the tympanic membrane is found on the side of the head; there is no protective outer ear canal as in birds and mammals. A straight bony rod in two sections (columella and extracolumella) provides a direct link between the relatively large tympanic membrane and the smaller oval window. The basic function is similar to that found in the anuran ear. The middle ear system in reptiles likewise acts as an amplifier to overcome much of the energy that would ordinarily be lost at an air-fluid interface. Energy distributed over a large area (tympanic membrane) is collected and focused on a much smaller area—the oval window of the cochlea. The round window membrane provides the pressure release, as in other land vertebrates.

The cochlea is divided into two parts by the basilar membrane, with the oval window on one side and the round window on the other. Consequently, when the fluid of the cochlea is stimulated via the sound-conducting pathway of the middle ear through the oval window, the basilar membrane is in the path of fluid movement and is free to move accordingly. The basilar membrane, or papilla, is probably the only acoustically sensitive tissue in the reptilian inner ear. The hair cells reside on this membrane and thus move with it in response to sound, but the tips of the sensory cilia on the hair-bearing end of these cells are embedded in other tissue, which is in turn firmly attached to the cochlear wall.

The principle of relative motion is evident in the inner ears of all vertebrates. Sound is received by a mechanical system, with one part either anchored or comparatively dense or heavy and the other quite free to move in response to the molecule or particle displacement produced by a source of sound or vibration. This relative motion acts as an energy source and sets the stage for the transformation of mechanical energy into the electrochemical activity of the nervous system. In the reptilian ear it is the relative motion between the body of the hair cell and the sensory hair itself. In the anuran ear, in comparison, the hair cells are firmly anchored, and the hairs are connected to tissue which is apparently unattached and free to move in the fluid pathway. The significance of these differences for hearing acuity in amphibians and reptiles is not altogether clear. However, the mode of action observed in the reptilian ear more closely resembles that seen in birds and mammals. It may be a more effective and sensitive detector of fluid movement, particularly with the basilar membrane and receptor cells lying across the path of particle displacement in the cochlea.

In the reptiles and birds we see the emergence and elongation of the cochlea; in mammals it actually becomes coiled. Figure 4.4 shows the structures of the inner ears of a lizard and a crocodile. In both, the three semicircular canals (the organs of balance) together with utricle and saccule can be seen in the superior part, while the cochlear duct is visible as an extension of the inferior part. The lengthening of the cochlea is obvious in the crocodiles. The increase in length of the duct is especially significant because it in turn increases the surface area available for acoustically sensitive tissue—receptor cells on the basilar membrane. Such changes may have important implications not only for enhanced sensitivity but for an extended frequency range of hearing and improved discrimination of acoustic signals. These developments continued in the mammals, where the evidence for their importance in hearing acuity is better documented and understood. Whereas lizards and crocodiles are both highly vocal, many reptiles

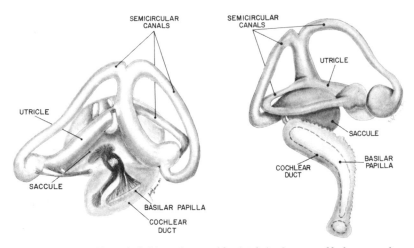

4.4   Inner ear of lizard (left) and crocodile (right). Organs of balance and auditory structures are indicated. (After Baird 1974.)

appear silent and their acoustic sensitivity is constrained to low-frequency sound such as might be transmitted through the ground or substrate.

Hearing in Reptiles

Evidence bearing on the auditory acuity of reptiles and on their use of signals in intraspecific communication is severely limited, due in large part to their relatively low metabolic rate and consequent low activity level, which renders them less than ideal subjects for behavioral study. With behavioral threshold data for only one species of turtle, it is necessary to turn to electrophysiological recording of stimulated electrical activity in the inner ear and auditory nerve in order to provide some indication of the nature of reptilian hearing. While such physiological measures are no substitute for behavioral measures which can reveal so many facets of the hearing process—as we have seen in fish and will observe to an even greater extent in birds and mammals—nonetheless they do give some indication of the frequency range of hearing and an estimate of relative acoustic sensitivity. An example drawn from the work of Wever is presented in Figure 4.5. The two functions in the figure represent recordings of the electrical activity of the inner ear in response to sound (cochlear microphonic response). The upper curve is from *Rhinophis,* a small burrowing snake with very limited low-frequency hearing, while the lower curve was obtained from a crocodile, which has a very sophisticated inner ear for a reptile—one which is, in fact, birdlike in nature. These two animals are representative of the extremes of reptile hearing, and their relative acoustic sensitivity is thus assessed on the basis of the inner ear's electrical activity when stimulated by sound. In this particular instance the relation between the ear's response to sound and the vocal output and ecology of the animal are noteworthy. *Rhinophis,* a silent animal that spends much of its time tunneling under the earth, is sensitive primarily to low-frequency sound produced by vibration of the ground. Crocodiles, on the other hand, are relatively noisy reptiles that vocalize and make use of sound regularly in intraspecific encounters. They are responsive to sound over a much wider frequency range (to 10 kHz) with greater sensitivity, and it comes as no surprise that their hearing, like the sophistication of their inner ear, probably rivals that of most avian species.

The one successful attempt to determine the actual hearing func-

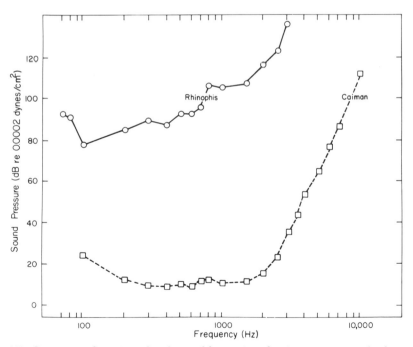

4.5 Sensitivity functions for the cochlear microphonic response in the burrowing snake *Rhinophis* and the crocodile *Caiman.* (After Wever 1978.)

tion of reptiles by means of a behavioral conditioning procedure was carried out by Wayne Patterson on one species of turtle. The technique was ingenious, in that it used the turtle's response of withdrawing its head into its shell as a measure of audibility. A line was connected to the animal's jaw permitting the experimenter to pull its head out from the shell. When the turtle's head remained out for several seconds, a conditioning trial was presented during which an auditory stimulus, a tone, was turned on and followed several seconds later by a brief, mild electric shock. The shock could easily be avoided if the turtle retracted its head in response to the tone, which the animal quickly learned to do. As a consequence of the conditioning process in which tone preceded shock, the tone served as a signal of impending shock, and the animal came to respond to it reliably in the same manner as it responded to the shock.

Following conditioning, tonal stimuli over an extended frequency range were reduced in intensity in an effort to determine their threshold level, and the results were plotted as shown in Figure 4.6. It is clear from these data that this species of turtle, at least, is restricted in

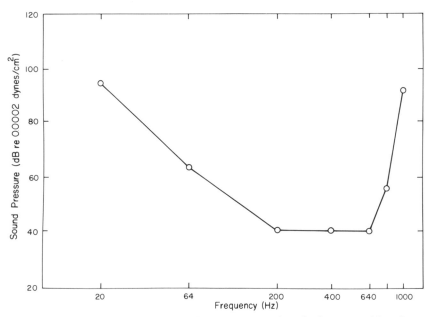

4.6 Auditory threshold functions for one species of turtle determined by a behavioral conditioning procedure. (After Patterson 1966.)

what it can hear, with regard to both sensitivity and frequency range. Its best hearing is between 200 and 600 Hz—in the lower part of man's frequency region for speech.

From the somewhat meager data on the hearing acuity of reptiles, on the electrophysiological response of their inner ear to sound, on their inner ear structure, and on their own use of sound in communication, it seems that the crocodiles and some of the lizards, such as the vociferous gecko, are acoustically the most acute of all the reptiles. Interestingly, most of these animals are nocturnal. Their use of sound and their sensitivity to it permit them to carry on an effective night life—a path not open to their distant, quieter, diurnal relatives. The remainder of the reptiles, including the snakes and turtles, are constrained in their hearing to a band of low frequencies below 1 kHz and have relatively poor sensitivity. In order for these animals to respond to a sound, it must either come from a source in their immediate vicinity or be unusually intense. The discriminative acuity of reptiles—that is, their ability to differentiate between sound frequencies or intensities—is unknown; but with the exception again of certain lizards and crocodiles, little in their behavioral repertoire would suggest the need for fine differential acuity. Many of the sounds that they appear to attend to, or even to make, include broad bands of frequencies and intensities with no obvious significance attached to slight changes in either of these acoustic features.

Communication in Reptiles

After a long period in which the study of communication in reptiles was largely anecdotal and the evidence difficult to confirm, the serious study of their vocal behavior and its significance is finally under way in earnest. The difficulties in examining these animals in the wild or in the laboratory are considerable, and a number of studies carried out under seminatural conditions (in zoos, on animal farms, and so on) have taken place with some success.

Sound emitted by many reptiles seems to serve as an antipredation device. Reptiles that are quiet under most other conditions emit a fairly broadband, intense sound when disturbed or threatened. The familiar hiss of most snakes and turtles comes to mind. Even the more vocal alligators emit a similar sound when threatened or disturbed. Croaking or squealing has been reported to fulfill a similar function in other reptiles. Since such calls are observed in so many reptilian species it is possible that deterring predators is one of the basic and primitive functions of acoustic communication in the reptiles as a group.

The lizards, such as the gecko, possess a larger repertoire of communication sounds, and these sounds serve a much greater range of functions in the behavior of these animals. The descriptive "multiple-chirp call" is shown in the sonographs in Figure 4.7 for three species of Mexican gecko. The call is pulsed and covers a range of frequencies extending from below 1 kHz to about 5 kHz, with a distinct chirping or clicking quality. For these geckos, there are clear differences in sound intensity (the darker the trace, the more intense the sound), in spectral composition, and even more clearly in chirp duration and in the interval between pulses. Although differences between individuals exist, there seem to be certain species-specific uniformities in acoustic structure of the calls which suggest the function of the call as a species-isolating mechanism. Such mechanisms are properties of species that hinder or prevent reproduction and thus genetic exchange with members of other species. The contexts in which these calls have been observed indicate that they are employed in agonistic encounters between males of the same species, perhaps in the settling of territorial disputes.

Single chirps, on the other hand, have been observed in lizards under conditions of distress and may facilitate escape from predation. "Churr" calls, which occur less frequently, are heard only in aggressive interactions between males and may serve as an acoustic threat. Calls may function in other ways in lizards, but until more careful observations are carried out under conditions as natural as possible, these ways will remain unknown.

Crocodilians (including the alligators) emit acoustic signals in many different contexts. Examples include their reaction to stressful conditions, vocal behavior associated with mating, intraspecific encounters, and so on; but unfortunately, observations of their acoustic behavior have been largely anecdotal and incomplete. More rigorous investigations of their behavior have failed to confirm many of these earlier reports. With an auditory system closely resembling that of most birds, we might expect from the crocodilians perhaps not bird song but at least something approaching the wide range and variation seen in the communication sounds of birds. In fact, these reptiles emit an astonishing diversity of sounds, vocal and nonvocal, as part of their communicative repertoire, but the specific function of many of these sounds is not known. As in other animal species, some sounds are context-dependent—that is, their function is specific to the circumstances in which they occur. Bellows or roars, hisses, growls, barks, coughs, grunts, and combinations of these all play a part in the social system of the crocodilians.

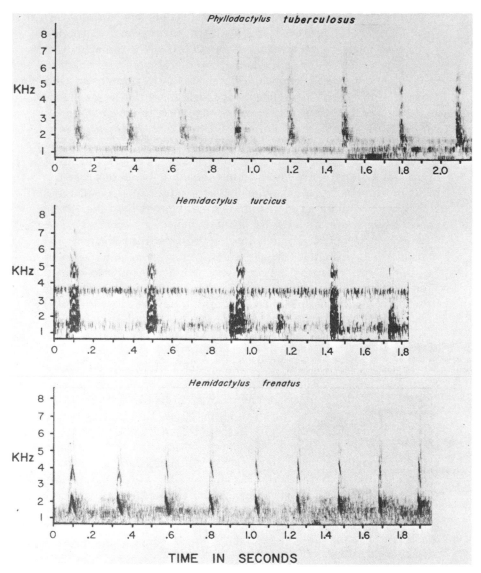

4.7 Sonographs of multiple-chirp call for three different species of gecko. (After Marcellini 1978.)

As an example, in the American alligator, bellows are of moderately short duration (one second) and consist mainly of low frequencies, with sound energy concentrated at about 200 Hz. At 3–6 meters from the calling animal, these calls attain a sound pressure level of as high as 90 dB and are easily detected through the air by humans at a distance of 150 meters. Such sounds are well suited for long-distance signaling and represent a successful adaptation to an environment where visual signals are difficult to see because of the heavy cover of vegetation. Bellowing may signal other members of the species and serve in the formation of breeding groups. The caller's sex and even individual identity is included within the signal. If the bellower is a male, the approach of a second male may result in an agonistic interaction, with a chase and subsequent fighting. On the other hand, should a female approach the vocalizing male, mating may take place.

A prominent nonvocal signal used by the American alligator is the "headslap," which entails slapping the head against the water's surface and which is transmitted effectively both above and below water. Frequently the headslapper is approached by conspecifics, but the function of the headslap is equivocal and it may be related to other events taking place at the same time. It has been observed during courtship, chasing, and bellowing.

Typical of the more recent findings reported above are those of Leslie Garrick and his colleagues at the New York Zoological Society. Such findings are particularly difficult to come by, for they involve painstaking hours of thorough and precise observations using behavioral sampling procedures in which observations are made at specific times of the day. These observations are recorded using tape, film, and special night viewing scopes. Acoustic events are recorded with directional microphones, and the data later analyzed by sonographic recording. Exhaustive descriptions of behavior and behavioral sequences and of the contexts in which these behaviors occur are essential. By means of techniques such as these, we are at last beginning to understand the functions of communication sounds in the shy descendants of that former great ruling class of vertebrates—the reptiles.

THE BIRDS

The reptiles provide a common ancestry for birds and mammals, although birds and mammals were descended from very different reptilian forms. In fact, because of their structural resemblance, birds are sometimes referred to as flying reptiles. The similarities in middle and inner ear morphology in birds and crocodiles further confirm that they are both the living remnants of the former ruling reptiles, the archosaurs, which included the great dinosaurs.

Most birds are, of course, characterized by their special adaptations for flight, and sight plays a significant role in their everyday affairs. Hearing is important particularly when birds rest or perch, and it serves a special function in communication over long distances in vegetative or arboreal environments where vision may be partially or completely occluded. Living birds are generally divided into about twenty-nine orders of which the Passeriformes (commonly passerines) are the most recently evolved. The passerines, or "perching birds," which represent more than half of all the living species of birds, are considered the most highly evolved avian order and include in their number all the many varieties of songbirds. The songbirds are of interest because of the complexity and significance of their acoustic communication system. There are nonpasserines such as the owls, who have evolved unique adaptations for highly sensitive hearing yet possess a relatively primitive communicative repertoire.

Auditory Structures in Birds

Birds and mammals are distinguished from other vertebrates by the presence of an external ear and an ear canal which leads to a recessed ear drum. Some have suggested that the canal serves a protective function. Perhaps more importantly, its tubelike shape allows it to resonate at certain frequencies, thus amplifying the sound energy of these frequencies at the ear drum or entrance to the middle ear. In many birds, the outer opening of the ear canal is surrounded by specialized feathers in the form of a funnel which acts as an efficient sound collector. It is likely that such structures provide an auditory advantage to the birds over the tetrapods in which the ear drum itself is external—at the body surface.

Many of the owls have evolved highly specialized and unique adaptations in outer ear structure which may improve their auditory acuity and provide a basis for their highly accurate localization of sound. The latter function plays a major role in prey detection for the night-hunting owls. Some species have highly developed, adjustable ear flaps which enable them to change the size and shape of their ear canal entrance, thus emphasizing certain sounds relative to others. In some species of owls the ears are asymmetrically placed on the sides of the head, a structural adaptation not seen in any other animal and

4.8   Front view of barn owl's head. Beak at midline divides face into two parabolic surfaces, which aid in sound reception.

one which aids in sound localization. Finally, in certain owls the facial ruff, composed of specialized feathers, is shaped either like a single parabola or as two parabolas side by side separated by the beak at the midline (see Figure 4.8). The parabola serves as a very effective sound-collecting device by funneling sound to the ear canal, thus presumably increasing the animal's acoustic sensitivity.

The lack of obvious differences in middle and inner ear morphology among birds is somewhat disconcerting in view of the wide variation in their use of sound. Whereas the songbirds emit a considerable array of complex signals for intraspecific communication, other birds apparently confine themselves to a few relatively undifferentiated calls. Either the fine differences in ear structure have eluded researchers, or the differences are to be found in the central nervous system, or, what may be more likely, the selective pressures imposed on the hearing of birds in the course of evolution were not solely directed at vocal communication. The hearing of birds may have developed to satisfy additional ecological requirements.

In many of its features the middle ear of birds closely resembles that of the crocodiles and lizards. The columellar apparatus, a long, slender, bony (sometimes partly cartilaginous) rod, connects the tympanic membrane to the oval window of the fluid-filled cochlear duct of the inner ear. The advantage gained by the ratio of the areas of the large tympanic membrane and small oval window provides sound amplification, as in the middle ear of all tetrapods. In addition, the angle between columellar shaft and tympanic membrane is less than ninety degrees, suggesting that the sound-produced movement of the tympanum-columellar complex is not simply pistonlike but may resemble the function of a lever. The mechanical advantage gained may provide amplification of the sound stimulus over and above that produced by the areal ratio of tympanic membrane to oval window. The effectiveness of such a system for sound transmission rivals that of the mammalian middle ear. As in the crocodilians, the two middle ear cavities in birds communicate anatomically with each other. The significance of this unique arrangement is still poorly understood, but there is a strong likelihood that it is a major factor in the differential response of the two ears to a sound source in the environment and the subsequent detection and location of that source by the animal.

In discussing the avian inner ear, little can be added to the earlier description of the reptile inner ear, particularly that of the crocodiles (see Figure 4.4). The cochlea, which lies within the bony walls of the skull, is a long, slightly bent tube which is more than 4 millimeters in length in the caiman and pigeon and over 10 millimeters in the barn owl. Its form and relations with other structures vary somewhat in different species. The basilar papilla, containing the sensory hair cells, extends over most of the length of the cochlear duct, while the lagena, a small patch of sensory tissue which may also serve an auditory function, is located at the inferior end. The hair cells on the basilar papilla are covered by the tectorial membrane in which their tips appear to be imbedded. As in the reptiles, it is probably the relative motion between the hairs and the hair cell bodies that is responsible for translating or decoding the acoustic message for electrochemical transmission within the auditory nerve to the auditory areas of the brain. Of interest for the functional relevance it may have to avian hearing are two structurally different types of receptor cells located on different portions of the basilar papilla. Only in birds and mammals do we find two such well differentiated types of receptor cells. Their significance in mammalian hearing is currently one of the most studied problems in auditory physiology.

The avian ear might reasonably be considered an advanced reptilian form, and yet the importance of sound to many species of birds, their vocal repertoire, and their perception of a variety of complex acoustic signals represent a significant difference in their acoustic behavior from that of most, if not all, of the reptiles. For the hearing acu-

ity of birds closely approaches that of the mammals over a limited frequency range.

Hearing in Birds

In contrast to reptiles, birds have proven able subjects in laboratory
experiments on their hearing. Thresholds for pure tone stimuli determined over a wide range of frequencies for at least sixteen species reveal a marked consistency. Birds differ as little in their sensitivity to
sound as they do in many of the characteristics of their peripheral auditory structures. A threshold function representative of all those birds
which have been tested is shown in Figure 4.9. Peak sensitivity occurs
at 2 kHz, which is clearly within the range of most bird song, although
the calls of some birds contain considerable energy as high as 8 kHz,
near the upper-frequency limit of their hearing. At the low-frequency

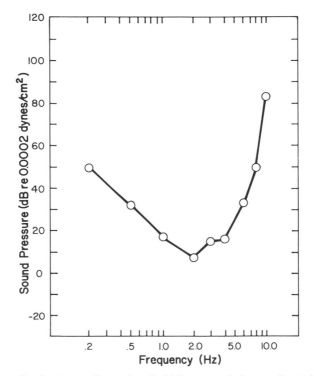

4.9 Generalized avian auditory threshold function (behavioral). (After Dooling 1980.)

end, their hearing extends to 200 Hz or below. Some very intriguing experiments on pigeons suggest a surprising sensitivity to infrasound as low as 0.5 Hz. Although it is clear that man's hearing does not extend to such low frequencies, these data from the pigeon have not yet been substantiated for other animals. Birds appear to be more sensitive than most amphibians or reptiles (caimans and geckos excepted), and their hearing extends to higher frequencies.

A variety of experimental procedures has been found effective for studying the hearing acuity of birds in the laboratory. They readily learn to respond to acoustic stimuli in order to avoid shock; they also can be conditioned to respond to the same stimuli when the response is immediately followed by a small bit of food (reinforcement) or to suppress their response for food when a sound stimulus precedes an unavoidable shock. In the first two situations the response, and in the third situation suppression of response, provide an index of hearing.

For example, a pigeon was trained in the chamber shown in Figure 4.10 in order that its threshold for sound detection and its frequency range of hearing could be determined. In this experimental arrange-

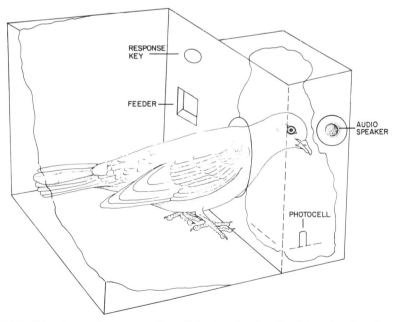

4.10   Side view and cutaway of conditioning chamber for determination of auditory sensitivity in the pigeon.

ment the bird was trained to make an "observing" response to turn a tone on, followed by a "reporting" response which indicated that it had heard the tone. Standing in the larger chamber, the pigeon learned to insert its head and neck through an opening into a smaller chamber affixed to the front wall of the first. A photocell and light arrangement in the smaller chamber registered movement of the bird's head through the light beam. At the same time its ear canals were directly in line with a small audio speaker attached to the side wall of the smaller chamber. The pigeon responded by moving its head in and out of the light beam (observing response), successively making and breaking the photocell circuit until a tone stimulus was presented through the speaker. The bird learned to report the tone by withdrawing its head from the smaller chamber and pecking with its beak a one-inch circular disc (reporting response) on the front wall of the larger chamber. Brief access to grain as a reinforcement immediately followed a correct response sequence. When the bird was well trained, the intensity of the stimulus was varied and the subject's threshold of detection for tones was determined over a wide range of acoustic frequencies. The results were very similar to those presented in Figure 4.9.

The concept of threshold in behavioral experiments as the minimum detectable level of stimulation to which an animal can respond is sometimes misconstrued as a fixed, immutable point above which the animal always hears the stimuli and below which it is always insensitive to them. In fact, nothing could be further from the truth. At relatively high intensity levels of stimulation the animal seldom fails to respond affirmatively, whereas at the other end of the continuum, at very low levels, it rarely hears the sound that is presented. In between these extremes there are sound levels which it hears sometimes and yet fails to hear on other occasions; in this region there is a limited range of stimulus intensities over which it displays a considerable degree of uncertainty. Testing at different intensity levels reveals that within this range there is a level which, on the average, the animal hears about half of the time; this is the threshold. It is a statistical entity, the averaged result of many determinations.

Although the threshold sensitivity function sets the limits of stimulation over which an animal is able to respond, other measures are essential to establish differential acuity within these limits. Bird song, for example, varies along many acoustic dimensions including frequency, intensity, and time. How finely are the various species of birds able to discriminate the differences that occur in the signals they hear? Amplitude modulation—the change in the intensity of acoustic stimulation over time—is an important feature of bird song and most bird calls. Different avian species appear to vary somewhat in their ability to resolve differences in sound intensity. For example, in the frequency range where the cowbird and canary are most sensitive, they can discriminate between tones which are separated by only 1.5 dB (in humans the intensity difference threshold is about 1.0 dB), whereas the pigeon and red-winged blackbird require a 3.0 dB difference for threshold resolution. In this instance, threshold is the difference in sound intensity that is correctly reported 50 percent of the time.

On the basis of the minute differences in the temporal structure of bird song, it was long thought that songbirds, at least, must have uncanny powers of temporal resolution—much better than our own. The supposition was based on man's inability to perceive the rapid frequency and intensity changes in the songs of certain birds. Recent behavioral data from the laboratory fail to support this assumption; they indicate that in the several species of birds that have been tested, there is little difference in temporal acuity between birds and humans. The question of temporal acuity is not a simple one, and the question may be asked or the experiment may be designed in several different ways. Subjects may be required simply to discriminate between tone durations, or, what is quite different, they may be asked to compare the lengths of a gap in an otherwise continuous signal. Gap detection thresholds in birds range from 3 to 5 milliseconds, compared to about 2 milliseconds in man. The results of other measures also confirm their similarity in temporal resolving power to man and other mammals.

One of the most salient characteristics of the vertebrate auditory system is its ability to resolve the small differences that exist in the frequency of most sounds. Like amplitude modulation, *frequency modulation*—that is, slow or rapid changes in acoustic frequency—is a common property of bird songs and calls. The frequency on which a song or call begins or ends or its dominant (fundamental) frequencies may signal an important function and, in fact, often enable a bird to recognize a member of its own species.

In the laboratory, birds are trained to respond to a change in the frequency of a pure tone and then tested in order to determine their *frequency difference threshold*—the change in frequency from the pure tone that they can detect on half the trials in which the tone was presented. Birds appear to be very sensitive to differences in frequency, and this sensitivity is similar across the several species that have been tested. At 500 Hz they are able to detect a frequency change of 10–15

Hz (compared to about 20 Hz for the goldfish and about 6 Hz for the macaque monkey). Between 1 and 2 kHz, where their hearing is most sensitive, their difference threshold varies from below 10 to slightly above 15 Hz. Based on frequency resolution of this order of magnitude, birds are able to extract substantial information regarding the frequency content of the acoustic signals they receive.

In birds, as in all animals, the ability to determine the exact location of a sound source may well be the result of one of the most persistent and intense selection pressures on the auditory system in the course of evolution. Sound localization depends on a variety of factors, including the structure or characteristics of the acoustic signal and the properties of the acoustic receptor organ, in addition to the integrating mechanism in the central nervous system. The circumstances surrounding the listener are also important, and one can appreciate the level of difficulty when both source and receiver are moving and the fix (or localization) must be carried out in three-dimensional space— a common situation for most flying birds.

At least in the land vertebrates, including the birds, it appears to be differential stimulation at the two ears with their openings on either side of the head that permits animals to obtain information regarding the spatial location of an acoustic event. In mammals, the two essential cues for sound localization are fairly well understood and will be considered in detail later (see Chapters 5 and 6). The mammalian head acts as a sound shadow at certain frequencies, so that the sound is more intense in the ear nearer the source. If it is a complex sound (containing many frequencies) it may also undergo changes in frequency composition between the two ears. At other frequencies, the slight delay in the time of arrival of a sound at the ear on the side of the head further from the source provides a temporal cue to the location of a sound. For example, a sound coming from the right side will enter the right ear shortly before it arrives at the left. In addition, the head can function as an obstacle by reflecting some of the sound, so that the listener perceives it as louder and perhaps of a slightly different pitch in the right ear than in the left. These two differences or disparities, one in loudness and pitch and the other in arrival time at the two ears, function as cues enabling the listener to tell very precisely whence the sound is coming.

In birds the situation is somewhat more equivocal and is perhaps complicated by the structural continuity between their two middle ears. Sound entering one ear can, with little loss in energy, stimulate the opposite ear through this middle ear cavity at the inner surface of the tympanic membrane. Differences in certain properties (intensity, phase) of the acoustic stimulus exist across the membrane, and these differences and their effects on the membrane are capable of providing information concerning the location of a sound source. Birds such as the owls may use these cues across the tympanic membrane and, in addition, because of their relatively large head size, benefit (as do the mammals) from the intensive and temporal cues provided by the head's sound shadow and the time delay across the head. The availability of so many cues to sound source location may help explain the uncanny ability of owls to locate their prey in complete darkness.

Countless observations leave little doubt that most birds are reasonably adept at locating the source of a sound and behaving accordingly. There is no question that most biological acoustic signals carry potential information about the location of the signal source, and that different avian species are morphologically specialized in varying degrees to receive this information. Unfortunately, there is little quantitative laboratory data to support these observations. Few measures of localization acuity have been obtained from birds, and our knowledge of the many factors involved in sound localization remains largely speculative.

By far the most complete account of directional hearing in birds is provided by the work of Eric Knudsen, Masakazu Konishi, Roger Payne, and their colleagues on the barn owl. The ability of these animals to locate their prey on the basis of acoustic cues alone offers a unique example of auditory adaptation. Infrared photography in a laboratory setting has revealed the barn owl's amazing skill in detecting the direction of movement of its prey in complete darkness. The owl then adjusts its strike pattern so that its talons are oriented along the long axis of the prey's body, thus increasing the chances of a successful capture. The secure insertion of both claws along the body's long axis also lessens the chances that the prey will be able to bite the owl's feet. Clearly, in terms of strike accuracy, the barn owl is a model of precision. Careful field observations of other species support these findings. The great grey owl, for example, is able to seize lemmings from deep snow without the aid of external signs or pathways on the surface of the snow; the sounds of chewing underneath the snow cover appear to be sufficient cues.

Accuracy, of course, is relative unless we can assign a number to it. In sound localization, the measure is usually taken in degrees of arc in either the horizontal (to the right or left of straight ahead) or vertical (up or down) dimension. The barn owl is capable of orienting to a sound source in space with an error of less than 2 degrees in either the horizontal or vertical dimension. This span of 2 degrees provides us with a measure of the owl's discriminative ability or acuity in localizing the source of a sound. When we use that figure for comparison with other animals, we find that the owl's acuity is greater than that of any terrestrial vertebrate tested, including man. Knudsen and Konishi, in carrying out this work, were able to take advantage of the natural response of the barn owl in orienting its head in order to face the location of a sound directly. Their experimental paradigm has a simple elegance; their findings offer a precise and reliable view of the barn owl's directional hearing and of some of the interesting underlying mechanisms responsible for it.

In the experiment the owl, with a magnetic search coil mounted on its head, was placed on a perch in complete darkness. Movement of the owl's head induced a current in the coil which was picked up by an amplifier. In this way the position of the head in space could be specified. Under the conditions of the experiment, two sounds were presented successively to the owl, which oriented to each of them in turn. The first sound always occurred directly in front of the animal and served to fix its head in the reference position for the search coil. The second sound could be presented in any one of a large set of locations, away from the reference location. The error of the bird's swift head orientation movement to the second sound could then be measured with considerable accuracy. Amazingly, the owls were able to orient to brief sounds that had ceased even before they had begun to move their heads. In some manner not yet understood, they could perform an almost instant calculation of the horizontal and vertical angular coordinates of the sound's position and then move their heads to face that position—all in a fraction of a second and with an error of less than two degrees.

On the basis of their experiments, Knudsen and Konishi were able to establish that the facial ruff of feathers and the external asymmetry of the two ears are significant anatomical features in localization, and that the differences in time of arrival and frequency content of the signal at the two ears provide critical information regarding the location of the sound source. Because owls may be the most acoustically specialized of all birds, there is little about their considerable acuity in sound localization that can be generalized to other avian species. Theirs is a unique adaptation brought about by very specialized ecological pressures related to prey detection in a nocturnal setting.

Communication in Birds

Bird song is continually in the air around us, and, because it resembles our own music, it has frequently played a role in our romantic literature. The scientific study of the sounds that birds make, which has contributed to our knowledge of their function in communication, has outpaced such studies in other animals. Current conceptual formulations regarding animal communication rely heavily on the results of observations and experiments on birds and their song. Their relative visibility and audibility, as well as their ubiquity, are undoubtedly the factors that make birds such popular and studied subjects.

The distinction between calls and song is perhaps more obvious in birds than it is in other animals. Calls often appear to be single elements of bird song. Conversely, songs tend to be longer and their acoustic structure more complex than calls. However, the calls of one taxonomic group may be more extensive than the songs of another. An example of the intricate structure of bird song is shown in the songs of three species of European warbler in Figure 4.11. Variations in intensity (darkness of trace), in frequency, in the duration of individual pulses and in the time between them, and in the spectral complexity (the frequency spread) are all evident. The differences among the songs of the three warblers supplied the primary evidence for their original separation into different species. Bird song clearly provides an important means of reproductive isolation for many birds.

Bird song is often, though not exclusively, the prerogative of the

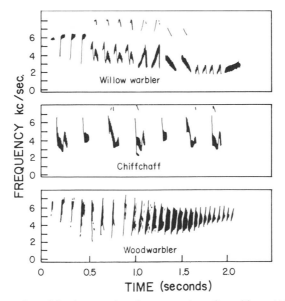

4.11 Sonographs of bird song for three species of warbler. (After Marler 1960.)

male, and is focused on reproductive activity. It is therefore seasonal and under hormonal control, aiding in the stimulation and synchronization of reproductive activity. Song in birds serves two functions. First, it promotes territorial defense: it announces ownership of a small surrounding area and serves notice to any intruders that they will be actively driven off if they approach too closely. Second, the song—or perhaps a slightly different variant of it—attracts eligible females for courtship and mating: it serves to establish and later maintain pair bonds, close relationships between male and female. In some species, duetting helps maintain contact between members of a pair and perhaps even synchronizes hourly behavior patterns.

Not only are there important individual and interspecific differences in bird song which are critical in recognition, but there is also geographic variation in the songs of the members of the same species. Zoologist Peter Marler and his colleagues, among others, have provided convincing evidence of regional dialects in different populations of the same species. These dialects are particularly evident from the songs of male white crown sparrows living along the California coast. Sonographs of birds from three different areas are shown in Figure 4.12. In the white crown sparrow, similarities in song structure among individual males in one region are such that even small differences between regions are immediately discernible. In the songs described in Figure 4.12, the differences in the second part of the song (in the right half of the figure) are most obvious. Animal behaviorists are still unsure of the function of dialects in animal behavior. They may help birds who have become widely scattered to rejoin their own group.

Bird calls differ widely and appear in a variety of contexts. As one example, the calls of two species of treecreeper are seen as sonographs in Figure 4.13. Calls *a–d* are used to indicate territories and to ward off other males of the same species. Calls similar to *c* and *d,* but shorter and with wider spacing between components, act as alarm calls when predators are sighted. If a bird is grabbed it will emit shrill distress calls resembling *e* and *f.* Calls *g–k* are used to maintain contact with immediate kin. Calls *l* and *m* are used on cold winter nights when the treecreepers, usually highly territorial, sleep together. At feeding time the adults use soft feeding calls (*n* and *o*), which apparently prompt the young to open their mouths. Calls *p* and *q* are begging calls, which the young utter as feeding time approaches.

Calls may be used to direct members of a species to a food source. They may act to synchronize hatching in certain species of chicks. In some species such as bank swallows, which live close together in colonies, calls function in individual recognition of young by their parents,

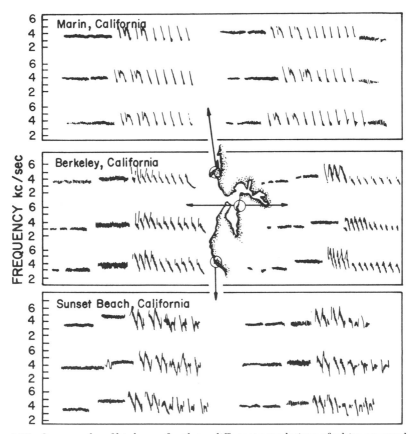

4.12 Sonographs of bird song for three different populations of white crowned sparrows in the San Francisco area. Geographic areas (Marin, Berkeley, Sunset Beach) are circled on the map, with arrows indicating the song for that area. Each song is about five seconds in length. (After Marler and Tamura 1964.)

particularly during feeding. Special calls are used at night by migratory birds and often before takeoff and landing. Some of the most unusual and highly specialized calls are those of the oilbird of South America and the cave swifts of southern Asia. Both species reside in dark caves and use the echo from their calls as a form of sonar for navigation—an adaptation found also in certain mammals, particularly some species of bats.

One of the most intriguing problems for study in the field of communication is the relationship between the acoustic characteristics of

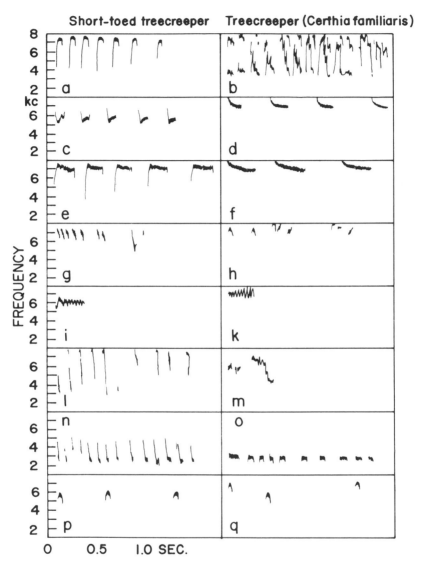

4.13 Sonographs of calls from two species of treecreeper. (After Thielcke 1976.)

a signal and its communicative function. The many ways in which a signal can be varied have important implications for the ease with which one signal can be discriminated from another, for the transmission of information over long or short distances, and for the accuracy with which a signal source can be located in space. Marler has suggested that birds giving alarm calls in the presence of a predator may reduce considerably the effectiveness of the principal cues for sound localization (for example, the differences in sound intensity and arrival time at the two ears) by selecting acoustic frequencies at which these cues are almost inoperable. Such sounds have a gradual onset and offset, are fairly brief, and are tonal in quality. The result is a high, thin whistle which is almost ventriloquial in nature. The matter, however, may not be so simple. Other researchers have argued that many alarm calls, by their structure, are fairly easy to localize; the caller directs attention to itself so that the predator may know that it has been spotted and so that the caller will be a difficult quarry. Which theory is correct may well depend on the particular species and on the circumstances in which danger threatens. The issue is not settled, but it serves to highlight the significance of the affinity between signal structure and communicative function.

Of all the vertebrates, the birds and mammals are the most highly vocal and make the most use of sound in their daily encounters. Acute hearing has provided birds with an important advantage for their particular style of living above the ground. Excellent vision is a characteristic that most diurnal birds rely on, but there are many environments where sight may often be occluded and thus not always dependable. Sound provides a useful and efficient means of transmitting information to kin and other species members under these circumstances, and birds have exploited its advantages with a high degree of success.

5

THE TERRESTRIAL

MAMMALS

IN THEIR EVOLUTION as a vertebrate class during the Age of the Reptiles more than 70 million years ago, the mammals owed a generous measure of their early and later success to their sense of hearing. Radiation of the earliest mammals into nocturnal niches required less reliance on vision, which their reptilian forebears had developed moderately well, and more on their senses of smell and hearing for guidance in their environment and later for communication. So rich and so diversified are the mammalian adaptations for hearing that they have been accorded three chapters in this book. The first, focusing on the terrestrial mammals, treats the mammals generally from the point of view of their evolution and their auditory characteristics. Chapter 6 considers the unique adaptations of the aerial and aquatic mammals (bats and cetacea, respectively), particularly with regard to their use of echolocation for navigation in their habitats. Chapter 7 examines the hearing of the primates. It is not for anthropocentric reasons that they take up an entire chapter, but rather because they have carried communicative skills a step further and because more is known about them than about other mammals.

EVOLUTION AND DIVERSITY OF THE MAMMALS

Although the first true mammals appeared some 70 million years ago, the evolution of the mammalian prototype started about 100 million years earlier, shortly after the reptiles began their ascent to power.

The group of reptiles from which the mammals arose split off from the ancestors of modern reptiles early in the evolution of the reptilian order. Consequently, any resemblances in auditory form and function between living mammals and reptiles would either have to hark back to their common ancestry almost 200 million years ago or be an example of convergent evolution.

There is little doubt that the mammals as a class have capitalized on their acoustic sense more than any other vertebrate or invertebrate group in the course of evolution. Their extensive use of hearing in all imaginable walks of life and the enormous diversity in their acoustic capabilities have increasingly become the subject of scientific research, and only recently have we begun to comprehend just what successful listeners the mammals are. How did all of this come about? What were some of the selective pressures acting upon the early and now extinct mammals that might have led to the many and varied auditory adaptations that are observed in their living descendants? In searching for an answer to such questions the facts prove elusive. Historical questions of this sort invite a certain amount of speculation which must be resisted and only partially succumbed to. The evidence, such as it is, is offered by paleontologists and consists of fossil material in the form of a few bones and teeth and perhaps fossil remnants of early plant life. The interpretation of this evidence is paramount to a consideration of the stresses and strains imposed on the early mammals by a hostile environment.

We are told that our mammalian and premammalian ancestors may have found some freedom from the huge predacious reptiles of that era by exploiting the night life. Such adaptations as the internal regulation of body temperature and bodily hair as insulation permitted better heat retention after dark, thus helping to support these animals in their nocturnal life style. At the same time, there is fossil evidence for changes occurring in the middle ear of those reptiles ancestral to the mammals—developments which continued in the earliest mammals. Apparently (for this is still somewhat speculative), there were changes in diet and metabolism which contributed directly to the evolution of the mammalian three-bone middle ear system. Increased activity level, metabolism, and voracity put considerable pressure on the development of improved jaw mechanics. The appearance of canines and a more powerful and efficient jaw articulation led in time to a foreshortening of the jaw, thus freeing from masticatory duty those bones in the posterior part of the jaw. Through a long and tortuous series of stages they eventually became the ubiquitous three-bone ossicular chain, a defining property of the mammals.

The evolution of the middle ear is well documented simply because bone is the substance of fossil remains. It is known that extensive changes were occurring throughout mammalian evolution in the inner ear and central nervous system, but the time course and systematic progression of these changes are impossible to substantiate because of the failure of soft tissue to leave a permanent record. The end results are, of course, observed in the living mammals. The living reptiles are our best candidates, and poor ones at that, for models of what acoustic structure and function may have been like in our premammalian ancestors. The resemblance is hazy because of the split which occurred in mammalian and reptilian evolution so many hundreds of millions of years ago.

The picture of the early mammals that emerges is one of a progressive series of many complex adaptations that were to carry these animals successfully through the reign of the great reptiles and well beyond. They hunted and foraged primarily at night and were small—in fact, shrew-sized—which enabled them to fit snugly into trees or underground burrows during the day when the large reptiles were afoot. Although their visual sensitivity at night was probably fair, they lacked color vision. They developed an almost uncanny sense of smell, and that, together with improvements in their hearing, guided their movements at night and permitted them at least some rudimentary form of communication with other members of their species. During mammalian evolution there were countless other adaptations (changes in brain size, for example), which were, like other adaptive changes, interdependent. Inasmuch as any one system can be teased apart from all the others that make up the whole animal, our concern is with hearing and acoustic communication and with the other changes and developments in structures and habits which may have affected the acoustic sense of mammals.

The diversity of the living mammals is reflected in some sixteen orders occupying almost every conceivable habitat—in the sea, in the air, and on land. Most current mammals share certain characteristics, such as the ability to internally regulate their body temperature, their nourishment of unborn young within the mother's body and their care after birth, an increased activity level, and increase in the size and complexity of the surface area of the brain (the cerebral cortex), and an enlargement of that part of the brain devoted to the sense of smell.

Compared to other vertebrates, all mammals have an unusually acute sense of hearing. Only the owl is a respected competitor, its ability to prey successfully on small mammals by auditory cues alone being its enviable hallmark. Many mammals, however, are able to respond to acoustic energy well below the hearing threshold for other animals. Their audible frequency range extends into the region of ul-trasound, far beyond that to which most other creatures are sensitive, owls included. Partly due to their extraordinary sensitivity and fre-quency range, mammals show a remarkable precision in their ability to localize the source of a sound in space—an adaptation that serves them well in foraging for food, searching for a mate, or eluding preda-tors. Selective pressures on social organization and acoustic communi-cation, similar to those imposed on the birds, have been at least partly responsible for the development of fine powers of discrimination—the capacity for resolving very small differences in the frequency, inten-sity, duration, and complexity of an acoustic signal.

Intimately related to these auditory capabilities are certain morpho-logical features that are found only in the mammals and that are very different from those observed in other vertebrates. The external ear appears for the first time in the terrestrial mammals and in the bats. A recessed tympanic membrane and three-bone ossicular chain form a highly efficient middle ear conducting system. In the inner ear is a unique coiled cochlear duct containing two distinct kinds of receptor cells. A highly complex central nervous system plays a significant but still poorly understood role in interacting with the auditory periph-ery—the middle and inner ear—and in exercising control over key auditory functions, such as the perception of communication sounds and sound localization.

Such are some of the more general auditory adaptations found in almost all mammals—at least in the marsupials and in the placental mammals. In the primitive egg-laying monotremes, such as the platy-pus and spiny anteater, both from Australia, many of these adapta-tions have not evolved to the stage in which they are found in the other mammals. These general adaptations will be considered in more detail below, with particular reference to those animals occupying ter-restrial habitats. Although many of these adaptations are shared by the bats, cetacea, and primates, later chapters will examine the highly specialized techniques employed by these animals in transmitting and receiving acoustic information.

AUDITORY STRUCTURES IN MAMMALS

For purposes of anatomical description, the mammalian ear is typi-cally divided into three parts or sections: the external or outer ear, the middle ear, and the inner ear. All are peculiarly mammalian. An ex-ample from the human auditory periphery is shown in Figure 5.1.

The outer ear consists of a skin-covered cartilaginous flap, or *pinna*

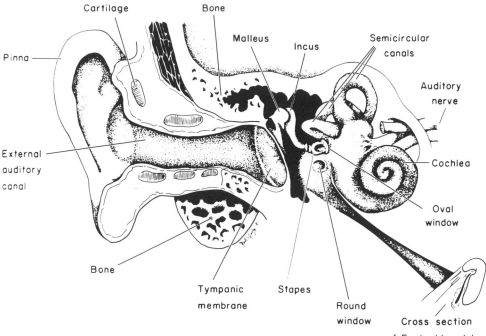

5.1 The human peripheral auditory organ, showing the outer, middle, and inner ear and the principal structures involved in sound reception and hearing.

(from the Latin word for feather), whose shape and size vary considerably among the many mammalian species. It is missing completely in many of the marine mammals but found in almost all others. Like the specialized feathers that surround the outer opening of the ear canal in birds, the pinna probably serves as an efficient collector and amplifier of sound.

Some animals with extremely acute hearing, such as the felids, or cats, exercise precise and differential muscular control over the movement and direction of the pinna. It is not at all unusual, for example, to see a cat orienting one pinna forward and the other toward the rear, enabling it to pay simultaneous attention to two distinct sound sources. Dogs appear to do almost as well by slightly elevating their pinnae ("pricking up their ears") in order, perhaps, to augment a faint but significant sound. Most higher primates, however, are unable to control their outer ear flaps—which, interestingly, is what most of us have in mind when we refer to the ear. Certain nonauditory functions have been suggested for the pinna. With its extensive vasculature, it most likely aids in the regulation of body temperature. At high temperatures, dilation of the blood vessels in the pinna allows the body to give off heat. In animals with pronounced and mobile pinnae (the elephant, for example) the external ear flaps, when moved forward so that they are at right angles to the head, may serve as an effective threat in both inter- and intraspecific encounters.

The pinna surrounds the tubelike external canal much like a funnel. The shape and extent of the canal vary in different species; it terminates at the tympanic membrane. As in the birds, the external canal resonates at certain frequencies, depending on its size and shape, thus enhancing some sounds relative to others. In humans, for example, the canal is resonant around 2,500 Hz, which is at the upper-frequency end of the speech range and in the most sensitive frequency region for human hearing.

The tympanic membrane represents the beginning of the mammalian middle ear (see Figure 5.1). Like the pinna, the middle ear, with its ossicular chain of three tiny bones, is strictly a mammalian invention. The *malleus* (hammer) is fastened to the inside or proximal surface of the tympanic membrane and articulates with the *incus* (anvil), which in turn is connected to the smallest bone in the body—the *stapes* (stirrup). The footplate of the stapes is fastened to the oval window of the cochlea, which forms the somewhat arbitrary boundary between the middle and inner ear. The mammalian middle ear is only roughly analogous to the columellar system found in other tetrapods. In both instances a similar function is performed: the transmission and amplification of sound from the ear drum to the oval window of the inner ear by way of a bony chain.

In mammals, as in the other tetrapods, the ratio of the area of the tympanic membrane to that of the oval window is large. Pressure spread over the relatively expansive tympanic membrane is brought to bear on the considerably smaller surface of the oval window, providing a major gain in sound pressure at the window. In addition, the arrangement of the malleus, incus, and stapes provides a lever action which adds additional amplification at the entrance window to the cochlea. By means of lever action and areal ratio, the ossicular chain of the middle ear restores most of the vibratory energy that would otherwise have been lost in the transfer from the air of the middle ear to the denser fluids of the inner ear. The efficiency of the avian middle ear system approaches that of the mammals at lower frequencies—to about 5 kHz; yet the mammalian middle ear achieves its major advantage at the higher frequencies where other vertebrates are unable to hear.

Although the basic morphological pattern and mode of operation of the middle ear are common to all mammals, there are unique variants which often reflect the selective pressures imposed by certain harsh and demanding environments. The work of Douglas Webster and other researchers on the kangaroo rat provides an illustration of one such unique adaptation. *Dipodomys,* the kangaroo rat, occupies a desert habitat, often living in underground burrows. Among its many adjustments to its environment are saltatory locomotion (jumping), water conservation, and a middle ear cavity, the internal dimensions of which exceed those of its entire cranial cavity. The characteristics of this large air space, together with a greater than usual amplification factor provided by the areal ratio and lever action of the three-bone ossicular chain, considerably enhance the conduction of lower frequencies (particularly those in the 1–3 kHz region) to the inner ear. Significantly, sounds produced by the kangaroo rat's chief predators, owls and snakes, are in this same frequency region. In experimental tests, surgical reduction in the size of the middle ear cavity rendered those kangaroo rats far more susceptible to predation than normal controls exposed to the same circumstances. The middle ear system of *Dipodomys,* then, represents a successful strategy, based on predator avoidance, for highly sensitive low-frequency hearing, since neither snakes nor owls are conspicuous for their noisy habits while hunting.

One of the most striking features of the mammalian auditory system is the *coiled cochlea* (snail shell) of the inner ear. In fact, the use of the word "cochlea" applied to nonmammalian vertebrates is inappropriate, for only in the mammals does this structure take on the appearance for which it was named. Only in the most primitive mammals—the monotremes—is the coiling of the cochlea incomplete; in the marsupials and placental mammals the number of turns ranges from two to five. By evolving into a tight coil, the cochlea considerably increased its length and surface area available for sensory tissue, without at the same time taking up much more room in the skull. A marked increase in cochlear length in the mammals appears to be related functionally to their ability to hear much higher frequencies than other vertebrates; however, within the various mammalian groups, differences in cochlear length and number of turns appear to be unrelated to an animal's audible frequency range.

Figure 5.1 shows the coiled mammalian cochlea, together with the outer and middle ear. The example shown is from a human ear. The cochlea contains three canals—vestibular, tympanic, and cochlear duct—and is shown in cross section in Figure 5.2. Although this example is from a guinea pig, it is representative of the mammalian form. The middle canal or cochlear duct contains the sensory cells for

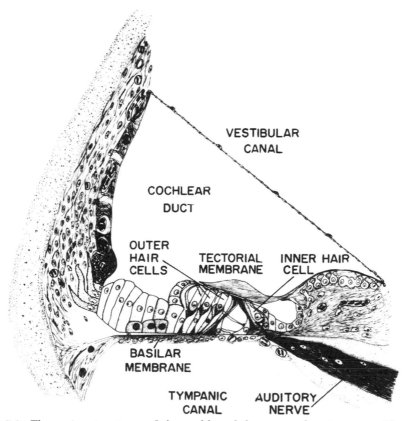

VESTIBULAR
CANAL

COCHLEAR
DUCT

OUTER
HAIR
CELLS

TECTORIAL
MEMBRANE

INNER HAIR
CELL

BASILAR
MEMBRANE

TYMPANIC
CANAL

AUDITORY
NERVE

5.2 The major structures of the cochlea of the mammalian inner ear. The cross-sectional view is of the cochlear duct of a guinea pig. (Courtesy of J. E. Hawkins, Jr.)

hearing, their supporting cells, and the individual fibers of the auditory nerve which contact the sensory cells directly. The *organ of Corti* is bounded above by the tectorial membrane—which rests on top of the hairs, or *stereocilia,* extending from the upper surface of the individual sensory cells—and below by the basilar membrane, which serves as a foundation for the sensory cells and their supporting cells.

One of the most renowned auditory scientists of the twentieth century was Nobel Prize recipient Georg von Békésy. On the basis of his early work, it is fairly well established that the motion of the stapes footplate in the oval window of the fluid-filled cochlea generates a traveling wave which proceeds up the cochlear spiral from its base at the oval window at the end of the vestibular canal to its apex, where the vestibular and tympanic canals join. The round window at the basal end of the tympanic canal provides the necessary pressure release. The fluid wave produces a relative motion between the sensory hair cells on the basilar membrane and the tectorial membrane. In some manner not yet understood, this mechanical stimulation triggers the sensory cells, which in turn transmit a chemical substance to the individual fibers of the auditory nerve, ensuring that the acoustic information will be sent to the central nervous system and brain.

In the mammalian organ of Corti there are two morphologically distinct types of sensory cells. A single row of inner hair cells together with three to five rows of outer hair cells follow the cochlear spiral from base to apex. Although the evidence is far from complete, it is likely that these two kinds of hair cells serve different functions. It has been suggested that the outer hair cells are the more sensitive and may account for the detection of levels of acoustic stimulation near threshold, whereas the inner hair cells take over at higher intensities and may, in addition, play a key role in frequency analysis.

One of the major goals of sensory physiology has been to decipher the code by which any sensory system receives energy from the environment and then translates it into the electrochemical language of nerve fibers. Individual nerve fibers, for example, code intensity of stimulation by altering the rate at which they fire or transmit energy along their length; an increase in stimulus intensity produces an increase in the firing rate of individual nerve fibers. One of the greatest challenges facing auditory physiologists has been to explain the vertebrate auditory system's ability to code acoustic frequency (what we often think of as the pitch of a sound). Considerably more attention has been paid to the mammalian auditory system, and this is evident in our substantially better understanding of that system.

Apparently, frequency is coded within the mammalian inner ear by two quite different mechanisms—one active for lower frequencies, the other for higher frequencies. At low frequencies (below about 3,000 Hz) nerve fibers fire or discharge in phase with the incoming acoustic stimulus. For example, a 500 Hz tone goes through a complete cycle or period every 2 milliseconds, a 1,000 Hz tone every millisecond, and so on (see Figure 1.1). The period of the acoustic stimulus is directly reflected in the periodicity of the discharging nerve fibers. At midrange and high frequencies, coding is executed by an acoustic frequency-to-place transformation. Less succinctly, the highest frequencies are registered by nerve fibers located deep in the base of the cochlea; successively lower frequencies are picked up by nerve fibers ascending in position within the cochlear spiral toward the apex. In effect, unrolling the cochlea reveals a map of frequencies, with high to low going from cochlear base to apex. These two conceptions of frequency coding in the mammalian cochlea have various names but are often referred to simply as the *periodicity principle* and the *place principle,* respectively. It seems fairly clear that the former operates in the coding of frequency in nonmammalian auditory systems. To what extent other relationships such as the place principle also function is unclear, although a crude approximation of the place principle exists in the ear of the locust (see Chapter 2) and has been well documented in birds.

Events taking place beyond the inner ear—in the central nervous system—exercise a significant degree of control over certain auditory functions. Although their exact role is unclear, nerve fibers directly transmit information, in both directions, between the central nervous system and the sensory hair cells in the organ of Corti. The business of locating the source of a sound and its direction of movement requires the two ears, and the process must be integrated in the nervous system where individual nerve fibers from the different ears pool the information they have obtained for transmission to higher levels in the brain. Clearly the nervous system plays an even greater integrative role when eyes must be directed to the newly found sound source and when the limbs are subsequently set in motion toward or away from its location. Although beyond the scope of this book, the importance of the brain and central nervous system in the further control and integration of all sensory information and its impact on behavior must be considered.

The mammalian ear as a structure represents a significant advance in the course of evolution. In every feature it is clearly distinct from the ears of other animals. The implications of these structural developments for hearing per se are just beginning to be understood; three of the most obvious are improved high-frequency hearing, greater sensitivity, and the enormous diversity in hearing acuity among mammalian species. There are almost no habitats left unoccupied by the mammals. The earliest mammals were terrestrial forms probably dwelling in wooded or bushy regions. Their migration to more exacting environments—the seas, the deserts, and the skies—relied in no small measure upon unique auditory adaptations enabling their survival and success over long periods of time extending to the present. The fact that terrestrial mammals remained in their original environment does not mean that they are, when compared with other mammals, less adventuresome in their life style or more conservative in the adjustments that have taken place in their hearing. In fact, among their number are the most sensitive ears the world has ever known. Their success over their competitors occurred on land, and thus there was little reason to seek a change in habitat.

HEARING IN THE TERRESTRIAL MAMMALS

Among the mammals the variation in auditory function is considerable. It is doubtful that mammalian hearing could be characterized by a single threshold curve, for example, as is possible for the birds (Figure 4.9). Such variation in acoustic form and function typifies the mammals as a group and must certainly reflect the very different adaptive strategies that have been selected for a range of habitats with substantially different requirements. The kangaroo rat, which must listen for the low frequencies in the wing beat of the owl and the rustling movement of the snake, provides an excellent example. Its middle ear structures and enormous middle ear cavity, as we have seen, represent an important part of its unique adaptation to its acoustic environment. The compelling evidence that these animals are able to respond effectively to the low frequencies in their environment is found in their threshold function, shown in Figure 5.3. The kangaroo rat demonstrates hearing which at the middle frequencies and above resembles that of many other mammals. The high degree of sensitivity and extended frequency range seen in Figure 5.3 are, in fact, uniquely mammalian. To this portrait the kangaroo rat has added unusually acute hearing at 125 Hz (C below middle C on our musical scale) and even below—a capability matched by only a very few other mammals.

Animal psychophysics encompasses the various techniques by which animals other than man are queried about their sensory experience. More specifically, it has provided the objective means by which we can determine the limits of sensory acuity and resolution in creatures without language. Whereas human psychophysics relies on language when instructing human subjects to attend and respond to certain changes in stimulation and not to others, animal psychophysics employs alternative strategies to achieve the same goal. The procedures, which are based on behavioral conditioning principles, have been successfully applied to many different mammalian species. The questions posed range from relatively simple ones regarding the minimum detectable level of energy to which an animal can respond (its threshold) to more complex functions such as the localization of a

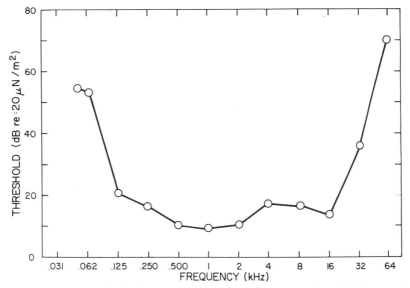

5.3 Auditory threshold function for the kangaroo rat. (After Heffner and Masterton 1980, and Webster and Webster 1972.)

sound in space or the discrimination of one signal from many with very similar properties.

The methods of animal psychophysics bring an animal to the point at which most human subjects enter an experiment. What verbal instructions accomplish for a human subject, behavioral conditioning procedures achieve for an animal. Through conditioning, the skills of listening or attending to specific stimuli—selecting some and rejecting others—are acquired. Once learned, these skills are put to the test; in psychophysical experiments, thresholds or some other measure of sensory acuity is determined. Animal psychophysics as a field is concerned with both method and data.

The guinea pig, a South American rodent and a domesticated laboratory animal and pet, provides an illustration of the effective use of conditioning techniques in animals for the purpose of testing hearing. A guinea pig is placed in a small test chamber (see Figure 5.4) within a larger, soundproof room. An audio speaker is positioned on the top of the chamber; a small push-button switch is recessed in the chamber floor directly beneath the speaker. The front end of the chamber contains a small trough to which small parsley-flavored food tablets (effective reinforcers for guinea pigs) are delivered from an automatic (electrically operated) food dispenser. The arrangement is designed

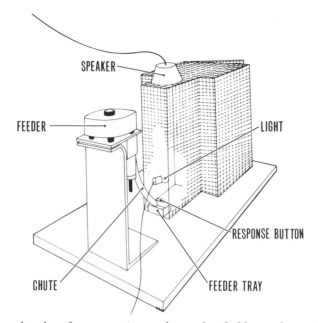

5.4 Test chamber for measuring auditory thresholds in the guinea pig. Speaker, response button, feeder with delivery chute and tray, and light are shown.

for what is known as *operant conditioning,* using food as a reward.

Initially a procedure referred to as shaping is employed. The animal has not eaten for a few hours, so that it arrives hungry at the test chamber. It is given food, at first, as a reward for simply approaching the floor switch. As it begins to spend more of its time near the switch the requirement for food reinforcement is gradually changed, and the animal goes through an orderly, predictable sequence of approaching, touching, and finally pressing and releasing the switch—the response which will be used subsequently to measure its hearing.

In the next stage of the experiment the requirement for food reinforcement is again changed, but gradually and through a series of steps. Stimuli are added to inform the animal about what to do next. In response to a flashing light the guinea pig performs the first response in the sequence—depressing the switch. The light stops flashing and becomes steady, telling the animal that it has made the correct response. The guinea pig continues to hold the button down until a tone is sounded. Upon hearing the tone, it promptly releases the button; the food dispenser discharges a food tablet, and the guinea pig moves toward the trough to retrieve its just reward. Mistakes such as letting go of the switch prematurely (before the tone is sounded) disrupt the sequence and cause food to be withheld, and the animal must endure a delay in the dark before the flashing light announces that a new trial has begun.

A trained guinea pig runs through this sequence smoothly with only an occasional error and is ready to have its hearing evaluated after about a two-month instructional period. Several stimulus intensity levels are selected, such that some will be clearly heard and others inaudible; thus, the range of stimuli will encompass or bracket the animal's threshold. These stimuli are then presented one at a time in random order and the animal's response is noted. Releasing the switch at the sound of the tone is tabulated as affirmative ("I hear it"), and keeping the switch depressed as negative (failure to hear). In the latter instance, if the animal remains on the switch, a different intensity level of stimulation is presented. The data are then graphed and make up what is referred to as a *psychometric function*. An example from one guinea pig is presented in Figure 5.5. The animal's

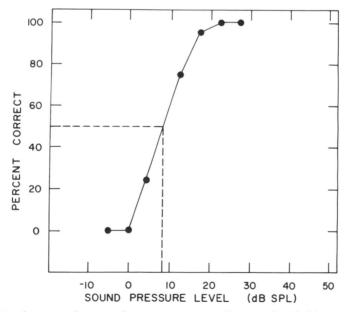

5.5 Psychometric function for one guinea pig. Hearing threshold at 16 kHz is determined from the 50 percent correct detection point shown on the function.

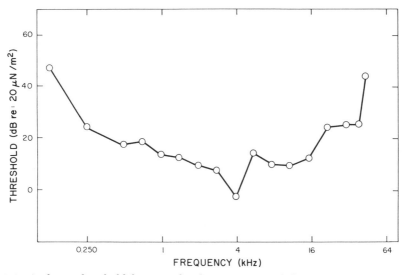

5.6   Auditory threshold function for the guinea pig. (After Prosen, Petersen, Moody, and Stebbins 1978.)

threshold for this single pure tone frequency is determined directly from the graphic function as that intensity of stimulation correctly reported on half of the trials (50 percent) in which it was presented. The calculation is shown on the figure. The same process is repeated over a wide range of frequencies, and the familiar threshold function that results is shown in Figure 5.6. In this manner, absolute acoustic sensitivity (the minimum level detectable) and frequency range of hearing are determined for the guinea pig. Although minor details may vary, the same procedures have been successfully applied to many different mammalian species.

The sensitivity of the guinea pig's ear and the frequency range that it encompasses are prototypically mammalian. The function represents something of a norm for modern terrestrial mammals with moderately good low-frequency hearing, maximum sensitivity between 4 and 8 kHz, and high-frequency hearing extending to about 60 kHz. Many terrestrial mammals, including some of the primates, approach this norm, which suggests that it satisfies the demands of many terrestrial habitats. Although such a statement must be considered conjectural, there are reasons why it should be considered. If the kangaroo rat is an exception, so is the cat, and, for both, acoustic adaptation to specific pressures from the environment has brought success. For the kangaroo rat, as we have seen, low-frequency hearing has aided substantially in escape from predation. For the carnivorous cat, with its nocturnal habits, an extraordinary sensitivity to sound over a wide range of frequencies has made hunting a highly successful venture. There are other examples, too, of living mammals that have responded to certain environmental demands by forging a life style in which exceptionally acute hearing serves them in one capacity or another; thus, although the guinea pig's threshold function may represent typical mammalian hearing, there is marked variation around that norm. Within the limits of acoustic frequency and intensity set by their threshold function, the mammals are able to discriminate an almost infinite variety of sounds on the basis of their difference in intensity, frequency, and duration. The distinction between prey and predator, the approximate distance and direction of movement of a remote acoustic disturbance, recognition of kin—all may depend wholly or in part upon the ability to distinguish differences in the pattern of energy reaching one or both ears.

Changes in acoustic frequency, sometimes slight and sometimes rapid, are almost always a property of biological signals, often in the context of communication. As in bird song, such shifts in the frequency of mammalian calls may reveal significant features of the signaler, such as its identity, mood, and certain aspects of the message being transmitted. In questioning an animal about its ability to resolve small differences in frequency, simple pure tones are used. A trained subject learns to report every instance of a change in frequency, and the limits of this ability are probed by varying the frequency difference until a *frequency difference threshold* is obtained. Like the detection threshold, this is defined as the frequency separation that the animal is able to report correctly 50 percent of the time.

Frequency resolution measured in this manner differs greatly among the mammals. That this difference is related in some way to the complexity of an animal's communication system is suspected although not yet known. A common feature of frequency difference thresholds in all animals is their orderly relation to the base or standard frequency at which they are determined. Difference thresholds increase as the frequency at which they are measured is increased. For example, the guinea pig is able to discriminate tones which differ by only about 12 Hz at 500 Hz, by 50 Hz at 1,000 Hz, and by 600 Hz at 10,000 Hz. In most, if not all animals, it is the low frequencies which carry the most information, and it is there that frequency resolution is the most acute.

Judgment of an acoustic signal's amplitude or intensity can be used to disclose information about the distance of a sound source from the listener. Signal intensity, like frequency, is an important cue in sound localization and may play a vital role in communication. The intensity of a vocal signal may often be directly related to the emotional state or arousal level of the caller. Intensity difference thresholds, determined in a very similar way to frequency difference thresholds, indicate a considerable discriminative acuity in mammals (as in birds) over an extended frequency range. Unlike frequency resolution, intensity discrimination changes little across an animal's normal frequency range of hearing. Guinea pigs, for example, are able to discriminate about a 3 dB difference between moderately intense stimuli over much of their frequency range. It is true, however, for all animals, that it is more difficult to differentiate between low-level stimuli (close to absolute detection threshold) on the basis of intensity differences.

Bruce Masterton and Henry Heffner, based on their research and the work of others with a variety of different mammals, have suggested that one of the most persistent selective pressures on mammalian hearing and on the mammalian auditory system has been related to sound localization. The high-frequency hearing observed in so many mammals may, they suggest, be an adaptation especially suited for accurate sound localization. Their ingenious argument is as follows.

The ability to locate the source of a sound at any distance depends very much on the difference in sound stimulation reaching the two ears. The simple fact of the spatial separation of the two ears means that a sound coming from any direction except straight ahead or straight behind must reach one ear before it reaches the other. The magnitude of this difference in time of arrival at the two ears, or Δt, will depend to a considerable extent on the space between the two ears or, more succinctly, on head size. The larger the head, the more pronounced will be this particular cue for localization. In addition, the head acts as an obstacle, casting a sound shadow. The size and effectiveness of the shadow likewise depend on head size. Both sound intensity and the spectral (frequency) composition of other than simple acoustic signals are affected: the head may reduce sound intensity and alter the spectral composition of an acoustic signal, thus producing differential stimulation at the two ears, or Δfi, a second cue for sound localization. Sounds of sufficiently low frequency that their long wavelengths approach the interaural (between-the-ears) distance are not shadowed by the head and fail to provide the Δfi cue. In effect, small heads could create a serious problem for precise sound localization. The time delay between the two ears may be so brief that the auditory system is unable to resolve it, and the head so small that it fails to cast a shadow except at very high frequencies. The key, then, to the Masterton and Heffner argument is that small heads must have ears which are sensitive to high frequencies in order that their owners may reliably locate in space those sounds which appeal and those which threaten. The temporal cue, Δt, may be unreliable; these small animals must presumably rely on Δfi, the intensive and spectral differences at the two ears.

The earliest mammals were diminutive, even tiny—shrew-sized. Small size was a successful adaptation to the many exigencies of their environment. Small animals moving about after dark were better able to avoid diurnal predaceous dinosaurs; in addition, they were more successful in finding shelter during the day because of their small size. The mammalian reproductive strategy of viviparity (bringing the young forth alive) and their ability to internally regulate their body temperature were related evolutionary developments. It seems likely that high-frequency hearing occurred with these other changes and that it was primarily an adaptation for accurate sound localization necessitated by small size.

COMMUNICATION IN TERRESTRIAL MAMMALS

Many mammals, like birds, have evolved complex and multifunctional acoustic communication systems. Social organization as a form of adaptation to the environment, related at least in part to procurement of food and predator avoidance, requires effective communication among members of the same species. How else can the many activities of group members—kin and conspecifics—be coordinated? Roger Peters, in his excellent book on function and meaning in mammalian communication, distinguishes four principal message systems, or categories of communication: neonatal, integrative, agonistic, and sexual. These are functional and pragmatic systems in which the message content or transmitted information is deduced from the message form, the context in which it occurs, and the behavior of the animal to which the message is directed.

The neonatal period refers to that relatively brief period after birth when the infant is under the close supervision of the mother and their interaction prepares the offspring for later more complex affiliative interactions. The acoustic signals are often given at a low sound level because the sender and receiver are in close proximity, and both are more vulnerable than usual to predators. The "mhrn" of the mother cat is a familiar example, as are the purrs of young kittens, which usually act as contact calls informing the mother that all is well and that no response on her part is necessary. In marked contrast are the distress calls of mammalian infants which are intense and solicit immediate attention by mother or other kin.

Rat pups emit an ultrasonic distress call in the frequency region of 32 kHz that invokes the mother to retrieve them when they have strayed or fallen from the nest. The frequency of the call precludes its being heard by predators whose hearing does not extend to such high frequencies. Further, since high frequencies attenuate so rapidly with distance, the pups' calls are heard only by those animals in the immediate vicinity—most often the mother. Many species employ more than one such call in their repertoire, indicating varying degrees of distress ranging in severity from hunger to actual seizure by a predator. The fawn, for example, adds an extra syllable or note to its call when the danger heightens. Infant shrews and beavers vary the intensity of their distress calls in direct relation to the degree of impending danger.

Integrative messages are those which bind together the social group and serve to strengthen its organization. Contact between group members, particularly when visual contact is broken, is facilitated by calls which may affirm the well-being of the caller in addition to fixing its distance from the listener and its approximate location. Sometimes the acoustic properties of the call are such that the caller's location remains cryptic—a strategy which aids in predator avoidance. Often contact calls are answered in kind by group members although no overt action may be taken. Shrews make a relatively soft twittering sound to maintain contact over short distances. Long-range contact calls are exemplified by the howling of wolves, which may lead to assembly by the pack, although in some instances it may serve to aid in establishing territorial boundaries.

Other forms of integrative messages include alarm calls. Given by many mammals, these calls are often intense and harsh—that is, noisy and composed of a wide band of frequencies, with the higher frequencies in the band containing more energy. Wolves, who fear only bears and humans, use only one alarm call: a short (0.1 second) bark, with the frequencies spread between 320 and 900 Hz. Such characteristics make the sound easy to localize. Barks may serve to terminate a howling session as a warning to other wolves to be quiet and not reveal their position, or as a message to wolf pups to return to their den if a predator has entered the neighborhood. Adult rats emit an intense chattering squeal which usually causes other rats to flee or hide.

Some deer snort or make a coughlike sound when the risk of danger is relatively low, but utter an intense high-frequency scream when wounded or attacked.

There is some question and considerable controversy about the function of alarm calls. According to the traditional notion, which has recently been questioned, the survival of the species (or other group) is often at stake, and the alarm caller, a true altruist, willingly lays down his life for the good of the species. Others dispute this, arguing convincingly that the caller, in sacrificing himself, is protecting his own genes, since these are carried by his kin who are those most likely to be close by—the receivers and beneficiaries of his alarm call. Still another argument suggests that the caller, by directing attention to himself, in fact avoids predation; if the others flee, he can choose the most secure position in the group. In the absence of sufficient data the debate continues.

Distress calls—which are expressed only under extreme conditions, such as when attack is imminent or actually in progress—are usually very intense sounds. It is when the snort or cough of the deer changes to a scream that distress is evident. The function of these calls is not altogether clear, although the way they act may be similar to the way alarm calls do.

Agonistic messages between conspecifics are often threatening but frequently serve to reduce the risk of conflict among group members which could result in fighting and even death. The agonistic message may lead to withdrawal by the receiver. Common threats in felids and canids range from snarling and growling to roaring. Hissing and spitting are particularly common in cats and may represent defensive rather than offensive threats. Growls and roars are intense sounds, spectrally broad, but with much of the energy at low frequencies. Hissing and spitting, on the other hand, are only moderately intense broad-band signals possessing considerable energy at high frequencies. Unfortunately, the relation between the acoustic structure of these calls and their function is not known. In contrast to growls and roars, hissing and spitting are not transmitted much beyond the immediate locality in which they occur. Growling and roaring, like piloerection (hair standing on end) and the open-mouth threat with the canines displayed, may act to increase the perceptual size of the threatening animals.

In socially organized animal groups, dominance hierarchies are established and maintained by threats. Tree shrews may employ a chirp call to reaffirm their rank. Rats in the course of threatening often emit ultrasonic pulse trains with pulse length between 3 and 65 milliseconds at frequencies in excess of 40 kHz. Submission by the receiver is often an effective response to threats. Rats commonly produce a high-frequency chatter around 26 kHz for several seconds or peep or whistle when another rat approaches too closely. Frequently, these calls will deter further hostility and the aggressor will leave the scene. If this does not happen, the peeping increases in intensity until it finally becomes an alarm or distress squeal. Should the aggressor continue his advance, the defensive animal will often attack.

As another form of expression of agonistic behavior, animals defend the territories in which they spend most of their time. Pikas, animals which live at high altitudes and which are small relatives of rabbits, appear to advertise their territory by the use of a short call sounding like their "ank" alarm call but apparently set off by the presence of a conspecific intruder on the resident's home ground. The prairie dog reacts to the same situation by barking. However, if it should bark on other than its own territory, it may be attacked. As mentioned above, howling by wolves may indicate a form of territorial behavior, by reemphasizing territorial boundaries and maintaining spacing between packs. Howling can apparently be heard by wolves at distances approaching, and in some instances even exceeding, the limits of their territories. Figure 5.7 shows a howling wolf pack and a sonograph of their unusual and, to us, plaintive call. The call shown in the sonograph is given by one individual.

Sexual messages center around courtship and mating and contain elements of both integrative and agonistic message systems. Affiliative and submissive behaviors, as well as those expressing dominance, are often observed. Sexual acoustic messages are less common and less frequently uttered by mammals than by birds. As in birds, these signals are usually given by the male as a means of revealing its presence and approximate location to estrous females and to other nearby males. To the females it might be considered a form of advertisement. Male pikas, in fact, engage in an acoustic display which contains the essential features of bird song (see Figure 5.8). The song is composed of a series of moderately complicated elements or calls which are repeated with a fundamental frequency of about 700 Hz. As the song continues, the individual elements are prolonged yet separated by increasing intervals of silence. When one male has finished another begins, until as many as four or five have made their presence known. This might be considered a form of rivalry among the males for the attention of the females. What determines the eventual selection of a mate is unknown.

In a similar vein, male elk engage in bugling. This unusual sound,

5.7   Wolf pack howling in full chorus, and sonograph of the howl from one animal. (Photograph by Roger Peters, courtesy of the Chicago Zoological Society; sonograph courtesy of Fred Harrington.)

rich in harmonics, shifts rapidly in frequency from 700 Hz to about 1,000 Hz, continuing for two to five seconds. Like the pika song, it is apparently a form of advertisement that provides essential information regarding the reproductive maturity and readiness of the male. Not an extremely intense call, it may be transmitted under optimal conditions up to several hundred meters. Bugling also occurs in aggressive encounters with other males and may serve as a defensive threat

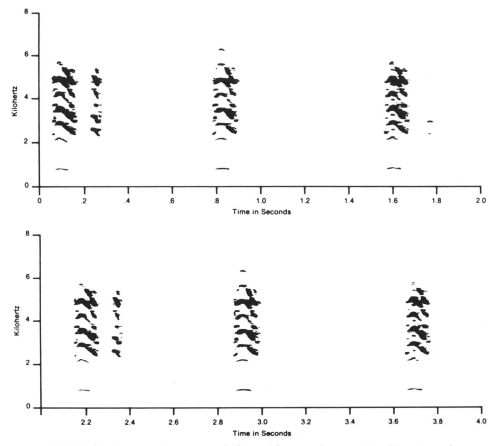

5.8   Male pika song in sonograph form (above and opposite). (Courtesy of Preston Somers.)

against the approach of nearby males who may be vying for the same females.

Peters' classification of mammalian communication signals into four types of messages provides a very helpful and systematic framework for the further analysis of mammalian communication and its relation to hearing and to the acoustic structure of the signal. It is a system that will be considered again in subsequent chapters when some of the more specialized mammalian auditory and communication systems are described. The mammals have the most highly evolved sense of hearing of any group. They are generally more sensitive to acoustic energy and more discriminating, and their hearing encompasses a far wider spectral range than other vertebrates. The ter-

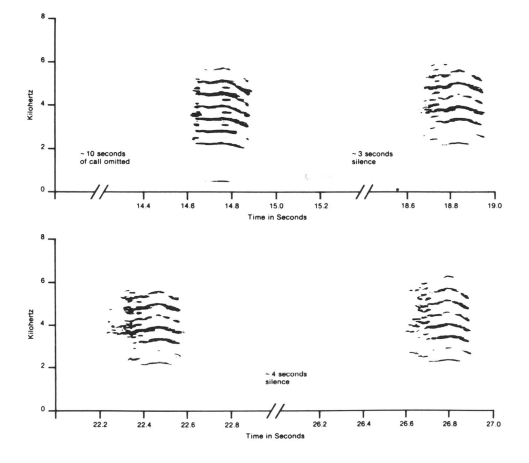

restrial mammals are no exception; yet compared to the songbirds, many of these ground-living mammals (most primates excepted) have a fairly modest repertoire of communication sounds. Much of their conversation, particularly at close range, is carried out in an olfactory mode. A powerful sense of smell, like hearing, was a successful adaptation of early mammals to nocturnal living. The few sounds that most terrestrial mammals produce are simple in acoustic structure and, for the most part, undemanding of such a highly developed auditory system. I would conjecture, in accord with Masterton and Heffner, that the finely tuned mammalian hearing mechanism is more closely geared to the detection and accurate localization of the sounds of friends and foes than it is to communication.

6

AERIAL AND AQUATIC

MAMMALS

THIS CHAPTER is primarily about bats and dolphins—two groups of mammals that left the land for the air and sea respectively, and, in so doing, evolved an unusual acoustic guidance system for navigating around their environment and for detecting prey. *Echolocation* is a means of imaging the environment by the vocal emission of sound pulses which are reflected from surrounding objects. The ear, in interpreting the returning echo, is performing in a manner analogous to that of the vertebrate eye. It extracts not only distance information but also details of shape and texture, in addition to the velocity and direction of a moving object, with a degree of resolution approaching that of the vertebrate eye. Although both bats and dolphins are sighted, their environments may often be occluded, thus rendering their vision ineffective. Most bats are nocturnal and many species apparently have limited use of their eyes. Dolphins may encounter turbulence in deep water where visibility is limited. However, their vision is quite good and useful under appropriate conditions—that is, at short distances or above water.

Echolocation is an effective and successful strategy for direction finding in ecological niches where light is either dim or absent. It may be considered an example of convergent evolution in the bats and cetacea (whales and dolphins), which have evolved independently for many millions of years from early terrestrial mammals. Their use of vocally emitted, ultrasonic (above 20 kHz) pulses and their sensitivity and discriminative acuity for the correspondingly high frequency echoes are perhaps the only feature that these two very dissimilar creatures share. However, among animals that hear, their adaptation is unique and deserves separate and special consideration. There are a few other animals that make some use of an echolocation system—the oilbirds of South America, some Asian swifts, perhaps some species of shrews, and even the marine catfish—but none, apparently, with the use of such high frequencies and with the extraordinarily fine resolution of the bats and cetacea.

THE BATS

It is generally agreed that the bats evolved early in the history of the mammals from primitive insectivore stock, although fossil remains of intermediate forms between bats and terrestrial insectivores have not yet been discovered. Like many mammals, bats are crepuscular and nocturnal, but they are unique among mammals in being the only members of the class that truly fly. Bats, or chiroptera (from the Greek word meaning wing-handed), are divided into two suborders—the microchiroptera and megachiroptera. Different species occupy a wide range of habitats from the northernmost reaches of the temperate zones to the tropics, and, like their distant terrestrial relatives, live on widely varied diets including insects, flowers, fruit, fish, amphibians, birds and other bats, and even the blood of terrestrial mammals. The smaller bats, the microchiroptera, will be the ones of interest to us here, for they are the primary echolocaters, nocturnal, mostly insectivorous, and possessed of limited vision.

Auditory Structures in Bats

The bat's ear is unquestionably mammalian, resembling closely in structure and function the ears of terrestrial mammals: it has an outer, middle, and inner ear with two kinds of receptor cells. There are some morphological variations in the ears of bats that may help account for their spectacular auditory adaptations, particularly their high-frequency hearing, and these bear close scrutiny here.

The external ears (pinnae) of microchiropteran bats are one of their most visible features. Although the pinnae vary greatly in size and shape between species, their prominence and ubiquity among the bats underscore their functional significance in hearing. Figures 6.1 and 6.2 show two insect-eating microchiroptera—the first a New World vespertilionid bat, and the second an Old World rhinolophid. *Myotis* has a pronounced pinna with parallel ridges on its inner sur-

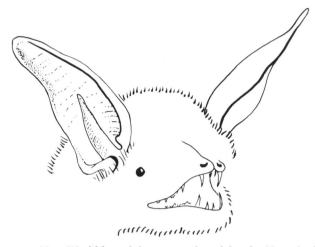

6.1 *Myotis,* a New World bat of the vespertilionid family. Note the large size and unusual shape of the pinnae. (After Sales and Pye 1974.)

face and a large tragus (standing in front of the pinna). As the animal scans its surroundings in order to pick up the returning echo from its vocally emitted acoustic pulses, the two funnel-shaped pinnae act as efficient sound collectors, and the tragus and ridges may possibly function as a wave guide for the important high frequencies. The external ear of *Myotis* has two other valuable features. The tip of the pinna can bend forward in response to intense ultrasound, producing about a 20 dB decrease in intensity at the inner ear for sound sources located directly in front of the head. And a small flap in the external canal leading to the ear drum can close off the canal, thus further reducing sound to the middle and inner ear. It is possible that these devices act in a protective mode, attenuating the bat's intense vocal emissions so that it might more clearly hear the returning echo. *Rhinolophus,* or the horseshoe bat (so named because of the shape of its noseleaf) lacks a tragus in its outer ear but exercises a considerable degree of control over the shape and orientation of the pinna. The two pinnae may be turned independently through wide angles and have been observed to vibrate rapidly and alternately. Such movements are apparently critical in echolocation; if they are prevented by surgical means, these bats become severely disoriented.

The bat's middle ear is distinguished by the relative size of some of its parts. The tympanic membrane is very small and extremely thin—an adaptation for high-frequency hearing. The middle ear cavity itself is small and uncomplicated, especially when compared with, say, that of the kangaroo rat (discussed in Chapter 5), which is an enormous cavity and very sensitive to low frequencies. Inside the middle ear the ossicles, particularly the stapes, are tiny, even for a small mammal. In contrast, the two middle ear muscles which attach directly to the ossicles are huge by mammalian standards. Their function, like that of the flexible pinna tip and mobile flap in the ear canal, seems to be protective. Just prior to or during the vocal emission of an ultrasonic pulse these muscles contract, thus reducing momentarily the sound-driven activity of the conducting ossicular chain and preventing temporary hearing loss as a consequence of the bat's own vocal output. The sensitivity of the inner ear for the returning echo is thereby preserved.

There is no mistaking the basic mammalian form of the bat's inner ear, yet there are some differences in detail and microstructure that seem related to the bat's use of ultrasound for echoranging. In bats, as in other mammals, the basal or lower portion of the basilar membrane near the oval window is implicated in the reception of the higher frequencies in the animal's audible range. Lower frequencies are perceived at progressively higher regions along the membrane. In some microchiropteran bats the length and/or thickness of membrane devoted to high frequencies is much greater than that found in other mammals. In certain rhinolophid bats a greatly expanded area of basilar membrane in the base of the cochlea is committed to a narrow

6.2 *Rhinolophus*, an Old World bat of the rhinolophid family. Configuration of the pinna, like that in *Myotis*, is important for sound reception. (After Sales and Pye 1974.)

band of frequencies between 82 and 86 kHz which contains the most significant information in the returning echo. The bat's basilar membrane and cochlea are not particularly long relative to those observed in other mammals. It is the morphological specializations in those basal areas sensitive to higher frequencies that seem to differentiate the bat's inner ear from the inner ear of other mammals.

Hearing in Bats

Zoologist Donald Griffin provided the first convincing evidence that bats rely upon the returning echoes from their own vocally emitted acoustic pulses in navigating and in hunting for food at night. It was Griffin, in fact, who first coined the term "echolocation." His extraordinary book *Listening in the Dark* recounts his early work on this problem in the late 1930s and early 1940s at Harvard. Although the bat's use of hearing rather than vision for guidance and prey detection had been hypothesized as early as the eighteenth century by the Italian scientist Lazzaro Spallanzani, it was not until the mid-twentieth century that the state of electronic technology made possible Griffin's discovery that bats are listening to the echoes produced by their own sounds, thereby accurately imaging their environment. Some of Griffin's early experiments will be described later in the chapter.

Griffin's results, in establishing the bat's use of a form of high-frequency biosonar, furnished powerful, although indirect, evidence that bats operate in an acoustic frequency region inaudible to man. In addition, bats possess excellent discriminative acuity for high frequencies as an effective alternative or substitute for sight. The behavioral experiments which confirmed the bat's ultrasonic hearing awaited the development of a suitable methodology for conditioning and animal psychophysics. The critical experiments were not carried out until some twenty years after Griffin had demonstrated the importance of ultrasound in the bat's hearing.

There was little doubt that bats could hear high frequencies. In addition to Griffin's work, physiologists had shown convincingly that the bat's inner ear and auditory nervous system were well adapted for the processing of high-frequency sound. Physiological recording of the electrical activity of the cochlear hair cells (the receptors) and of individual nerve cells had made this amply clear. But it remained for the behavioral scientists to tell us what bats actually hear.

Bats are not particularly well adapted for a laboratory existence. The behavioral techniques which had proven successful with some of the terrestrial mammals were not easily applied to an aerial mammal that catches its food on the wing. One of the earliest successful attempts to condition bats in order to test their hearing was carried out by John Dalland in Glen Wever's laboratory at Princeton. The experiment was a tour de force and represented a breakthrough in the quantitative study of bat hearing. The subjects, New World vespertilionid bats of the genus *Myotis* (the little brown bat) and *Eptesicus* (the big brown bat), were trained to stand on a small listening platform facing a sound source or speaker. In the correct position the animal interrupted a light beam to a photocell which started a trial by triggering an electronic timing circuit. If the animal continued to maintain his stance on the platform, a pure tone signal was sounded briefly. The bat then acknowledged hearing the signal by leaving the platform and walking across the floor of its small test cage toward a food cup on the opposite side. It passed through a second photobeam on its trip across the cage, and, if it had correctly reported the tone, a live mealworm was delivered to the cup. If the report was incorrect the cup remained empty. In either event the bat then returned to the listening platform for the next trial.

Auditory thresholds were determined by beginning with a moderately intense, clearly audible tone and then lowering its intensity after each correct detection until the subject failed to leave the platform, thus indicating that it had not heard the tone. Again the tone was increased in intensity and the sequence of decreasing tone intensities was repeated. The threshold was based on a number of such passes and represented a point midway between the least intense tone reported by the bat and the next lowest intensity for which the animal remained in position on the platform. Threshold functions for *Myotis* and *Eptesicus* are shown in Figure 6.3. Extremely sensitive high-frequency hearing is clearly substantiated for these two microchiropteran species. Compared to most other mammals their low-frequency hearing is quite poor, yet they seem well adapted to receive the echoes from the high-frequency pulses they emit.

Careful scrutiny of the major bumps in the threshold function for *Eptesicus* reveals two frequency regions of high sensitivity—one, fairly broad, centered around 20 kHz, the other, much narrower, near 60 kHz. The band of frequencies around 60 kHz consists of those employed in echolocation, whereas those closer to 20 kHz may be important in sound communication among bats of the same species. Another, more complicated example of such division of function is seen in Figure 6.4. In this audiogram for the greater horseshoe bat, *Rhinolophus ferrumequinum,* three district frequency regions of sensitivity appear. The two lower, broad areas around 20 and 60 kHz are very

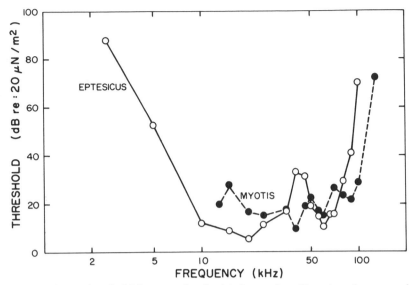

6.3   Auditory threshold function for the big brown bat, *Eptesicus fuscus*, and the little brown bat, *Myotis lucifigus*. (After Dalland 1965.)

close in frequency to those found in the audiogram for *Eptesicus*. The third area is centered at a much higher frequency and is much narrower, or, in the language of the bioacoustician, more sharply tuned.

The different valleys and troughs in the audiograms cannot be passed off as behavioral variability or measurement error. On the contrary, behavioral techniques and acoustic measurements are both extremely precise in these experiments. The graphic bumps represent real phenomena which may be appreciated if we consider briefly here (more fully later) the nature of the signals to which different bats are listening. The brown bat emits a frequency-modulated (changing frequency) pulse which sweeps in a downward direction across this animal's high-frequency region of sensitivity. The horseshoe bat, on the other hand, while echolocating, emits a slightly more complex signal—a single or constant frequency (CF) terminating in a brief downward frequency-modulated (FM) pulse.

The two components CF and FM elicit, as we shall see, quite different information from the environment. The constant-frequency component is just over 80 kHz and thus fits that high, sharply tuned area on the horseshoe bat's audiogram. The FM portion of the signal most likely corresponds to the middle sensitivity peak in the audiogram, whereas the broadly tuned area around 20 kHz may, as in the brown bat, represent frequencies employed in intraspecific communication.

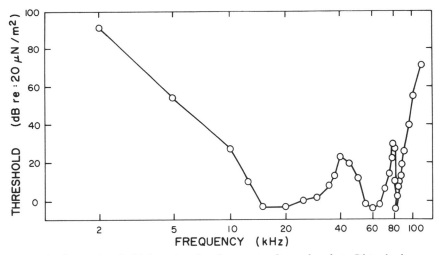

6.4 Auditory threshold function for the greater horseshoe bat, *Rhinolophus ferrumequinum*. (After Long and Schnitzler 1975.)

At least in the greater horseshoe bat, there is convincing evidence linking the finely tuned area of high-frequency sensitivity in the audiogram to the enlarged portion of the basilar membrane deep in the base of this animal's cochlea. To complete the picture, physiological recordings of electrical activity in response to acoustic stimulation in the cochlea and auditory nerve show the same sharp tuning just above 80 kHz. Clearly the inner ear is implicated in the highly specific sensitivity of the animal's hearing in this frequency region. The particular role that this narrow band of high frequencies plays in the horseshoe bat's echoranging behavior will become apparent later.

Communication in Bats

The extensive study of echolocation in bats has completely overshadowed the investigation of their use of sound in intraspecific communication, and for good reason. Their night life and their all but inaccessible communal roosts have made fruitful observation of these animals in their habitats almost impossible. Fran Porter's recent studies of captive colonies under seminatural conditions in restricted indoor areas have revealed some interesting aspects of their acoustic communication system which provide important leads for further study in the wild. On the basis of their cries, infant bats are recognized by their mothers, who may return them to their familial roost if they are found outside it. It is possible also that adults recognize each other at least partly on the basis of acoustic cues, but this needs further confirmation.

In contrast to their echoranging signals, the communication calls of bats tend to be longer in duration, and often, though not always, there is more energy at lower frequencies—below 30 kHz. Screeches and wideband strident and intense buzzlike calls are used by adult males in rounding up members of their harem who may have strayed from the roost area. These same calls, which serve an integrative function in one context, may also be employed in agonistic encounters by males chasing or attacking other males, often in an effort to expel the intruders from the home roost area. It is likely that these same calls are used in distress, since they were found to be emitted when bats were captured or handled by the investigators. Vocalizations by bats are also observed before and during mating. Frequency-modulated signals that are more slowly modulated than those reserved for echoranging have been recorded just before, after, and during the male-female pursuit that may precede mating. If the results of these indoor studies can be applied to bats under more natural conditions, we will probably find that these animals possess a fairly rich and complex acoustic repertoire for maintaining an orderly social existence, in addition to an echolocation system for orientation and guidance.

THE CETACEA

The evolution of the cetacea (whales and dolphins), like that of the bats, is unclear, although it is generally believed that they descended from very early carnivorous mammals. Also like the bats, the cetacea, in adapting to their new environment, underwent dramatic changes in form and function; they are perhaps the most atypical of the mammals. The two groups of living cetacea are the toothed whales, or odontocetes, and the whalebone whales, or mysticetes. Most living cetacea are odontocetes—the group that includes the dolphins and that lives primarily on a diet of fish. This group will be of greatest concern to us because, like the microchiropteran bats, the odontocetes employ biosonar for acoustic guidance and prey detection. Within the odontocetes it is the dolphins, particularly the Atlantic bottlenose dolphin, that have occupied the attention of scientists and that are the best understood of this cetacean group.

Auditory Structures in Dolphins

The problem in sound detection confronting the earliest cetaceans was the converse of that which greeted the ancestral tetrapods as they first ventured onto land. An ear which had been modified successfully for life on earth in the premammals might have needed a second revision if the cetacean mammals were to use sound effectively underwater.

Cetaceans have no pinnae; their very narrow and tortuous external canal leads to a relatively thickened tympanic membrane. The pathway for conducting sound to the middle and inner ear is uncertain. Although it has been suggested that the conduction route is the same as in terrestrial mammals, a more widely held view is that underwater sound is conducted to the inner ear through the body tissues located in the region of the external canal and in the lower jaw, and thence to the *bulla*—the bony cavity containing the cochlea. The effect on the inner ear is approximately the same as it would be if the conventional conduction route, external canal–tympanic membrane–ossicular chain, had been employed.

As in the bats, certain structural variations in the dolphin cochlea seem particularly well suited for sensitive high-frequency hearing. Although the number of receptor cells is about the same as in the primates, there is a much more extensive innervation by auditory nerve fibers, which may serve to increase the amount of high-frequency information transmitted to the brain and nervous system and to ensure the animal's considerable discriminative acuity. Detection and discrimination of high frequencies may also be enhanced in dolphins by a basilar membrane which is not only stiffer but varies considerably in stiffness along its length.

Hearing in Dolphins

Like bats, the odontocetes rely upon the high-frequency echoes from their own vocally emitted acoustic pulses for navigation, orientation, and hunting for food. It has been well known since the time of Aristotle that dolphins have an abundant vocabulary, since many of their sounds lie within the relatively narrow frequency limits of human hearing; but it was not until after 1950 that their use of biosonar was established. Winthrop Kellogg's interesting book *Porpoises and Sonar* describes many of these early investigations in a form and style suitable for the nonspecialist.

Evidence for the dolphin's sensitivity to ultrasound was beginning to unfold in the 1960s at about the same time that the earliest successful threshold measurements were being carried out in bats. Unlike bats, dolphins were proving very adaptable to experimental inquiry, which included at least some form of temporary confinement in large underwater pens and the use of operant conditioning with fresh fish as reinforcement. The intelligence of these animals is well known; having their hearing tested posed no serious challenge. If there were problems, they were more apt to occur in the accurate specification and measurement of sound underwater. In one of the standard procedures, an animal was required to position its rostrum (or beak) in a small trough so that it was pointed precisely at a sound speaker directly ahead. When the dolphin detected a sound, it would swim to the other side of its pen and, striking a response paddle underwater, receive its fish. The animal would then return to the rostrum station and wait for the next audible signal; if it responded when no signal had been presented, it was punished by a time-out or brief delay in the experiment. The resemblance to Dalland's experiments with the bat's hearing is unmistakable.

The threshold function for the bottlenose dolphin, which is typical of many of the cetaceans, is shown in Figure 6.5. Similarities to the microchiropteran bat's hearing are immediately apparent. The dolphin's frequency range extends from below 1 kHz to over 100 kHz, with maximum sensitivity between 20 and 80 kHz. Its low-frequency hearing, like the bat's, is not particularly striking. Evidence for the sharp tuning observed in specific frequency regions of the bat's audiogram is not obvious from a close inspection of the dolphin's audibility function. The perturbations observed in the threshold curve near 10 and near 100 kHz are quite small and may reflect slight variability in the signal source or in the animal's behavior.

Fine discrimination of the target characteristics in the returning echo or of the subtleties in the communication sounds of conspecifics implies excellent differential acuity for small changes in the properties of a sound—in its frequency, intensity, duration, or timbre. Beyond merely detecting the presence of an acoustic signal or echo which indicates an object somewhere in the environment, discrimination of these various features of a sound permits its identification, provides detailed information about such object characteristics as shape, size, and even texture, and allows an animal to estimate not only the distance of an object but its location, direction of movement, and velocity. Finally, in the event that the signal is a communication sound, differential resolution of its properties may allow recognition of the caller and proper decoding of the message content.

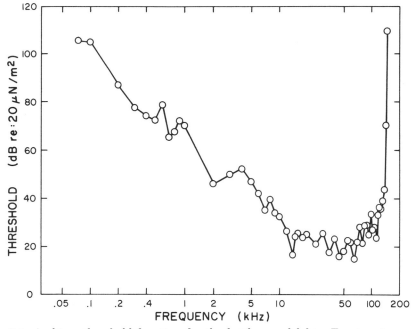

6.5 Auditory threshold function for the bottlenose dolphin, *Tursiops trun-catus*. (After Johnson 1966.)

The evidence that is accumulating indicates that dolphins, bats, and humans may be among the most highly specialized of all animals in their ability to resolve the differences in the fine structure of auditory signals. We infer these specializations in bats based on their ability to respond to minute changes in target characteristics in an echolocation task. In humans, such fine differential acuity is very likely related to our complicated acoustic communication system—human language. In the cetaceans, this adaptation for hearing the differences in various signal properties may be tied both to their echoranging and to their communication systems.

Difference thresholds for acoustic frequency in the dolphin confirm an extended sensitivity from 2 kHz to about 50 kHz, covering both the lower-frequency range for communication signaling and the higher frequencies used in echolocation. At 20 kHz, for example, a signal differing by less than 100 Hz is discriminable by the dolphin. Intensity difference thresholds of as little as 1 dB have been determined for the bottlenose dolphin, which can also discriminate sounds differing in duration by as little as 24 milliseconds. The dolphin's ability to localize the source of a sound underwater rivals that of bats and humans in air. Between 20 kHz and 90 kHz the angular resolution between two sound sources may be as little as 2–3 degrees for pure tones, and less than 1 degree for pulses of noise that contain a broad range of frequencies and that are similar to those used by the animal in echolocation. It is amply clear that the dolphin's hearing represents a superb and highly specialized adaptation for an acoustic way of life.

Communication in Dolphins

It is difficult in considering dolphins to separate myth from reality. With a large brain and cerebral cortex somewhat resembling man's, they are often credited with an intelligence which distinguishes them from other mammals and a language which rivals our own in complexity. Their history and habits are steeped in legend, and even today much of the evidence about their behavior is based on anecdote rather than scientific analysis. Their habitat and life style, like those of the bats, makes their study under natural conditions nearly impossible. What is known about their social behavior and communication comes mainly from the observation of small captive groups maintained in large tanks or aquariums. We must extrapolate from such data with care and healthy skepticism.

Dolphin communication sounds have been described as clicks and whistles; a third group are known variously and onomatopoetically as quacks, squawks, blats, or barks. Like the communication sounds of bats, those of dolphins tend to occupy a somewhat lower-frequency region than the clicks they use in echoranging. Whistles are among the most interesting and most richly diversified of dolphin calls. Investigators with underwater recording equipment have been able to differentiate many different whistles on the basis of their acoustic structure, leading some researchers to suggest that each animal has its own individual acoustic signature which is easily recognized by members of its own group.

Infant dolphins have been observed to whistle repeatedly when separated from their mother. As in other mammals under similar circumstances, the whistle may serve as a contact call. Whistles may also function as contact calls for adult dolphins and are often answered by others. Whistles and quacks are apparently used in soliciting attention. It is also possible that the dolphin's loud exhalation of air above the surface of the water may be used in reestablishing contact with other dolphins. It is not clear, however, that these exhalations are audible to other dolphins underwater.

Distress whistles are different in acoustic structure from contact whistles and may be emitted by lost infants or by dolphins that are sick or injured. The response by other dolphins is immediate, and rescuers return the call and quickly arrive on the scene. An animal that is incapacitated and has trouble in rising to the surface for air is often aided and actually lifted to the surface by its companions. Still another variant of the whistle is its function as an alarm call which may be emitted in the presence of known predators—killer whale, shark, or man. Dolphins are a garrulous lot, and their momentary silence may also signal alarm.

Agonistic behavior has not often been observed in dolphins. One form that such behavior takes is the jaw clap, which may represent a rather mild threat. The jaws are opened to show the teeth and then closed suddenly, producing a sharp, loud report. The gesture is often repeated and may be part of a whole complex of movements by the animal, including arching the back while facing the opponent as if the animal were preparing to charge. A whistle rising and falling in frequency twice, sometimes together with the tail flukes hitting the surface of the water, may reveal a stronger threat.

In concluding this brief survey of dolphin communication a word of caution is in order. Much of what has been reported here is limited to only one species, the bottlenose dolphin, and many of the observations were made under restricted (captive) conditions. It is known, for example, that certain species of dolphin do not whistle, but it is unclear what they substitute for this rich and diversified acoustic signal.

ECHOLOCATION

It is difficult for us to consider echolocation as a highly serviceable alternative strategy to sight. Our own experience with it seldom consists of anything more than projecting our own voice against a distant but visible hill or mountain and marveling at the acoustic fidelity of the returning echo. For the bat or cetacean, echolocation is far more than a single echo from one large object located at a considerable distance. The ear and auditory nervous system in the echoranging bat or dolphin, much like the eye and visual nervous system in other animals, are continually receiving information about the current state of the environment and about the many objects in it, near and far, from the echoes reflected from an almost infinite variety of stationary and moving surfaces. Before considering the echolocating bat or dolphin in action, it is helpful to examine the properties of the acoustic signal that these animals emit. The nature of that signal is an important part of the key to understanding the unusual and highly specialized auditory adaptation for biosonar in the microchiropteran bats and the odontocete cetaceans.

In a very real sense, acoustic echoes, in the information they provide, substitute for light reflected off myriad surfaces. Cues for object distance, size, shape, texture, and so on are present in reflected sound as in reflected light. Such cues are picked up by the auditory receptors, decoded and transmitted to the nervous system and brain, and quickly perceived by the animal, whose behavior takes account of their presence and relative importance to its livelihood.

Sound waves encountering an object will be reflected if their wavelengths are shorter than the object's dimensions. Consequently, higher-frequency signals with short wavelengths are ideal for detecting echoes from insects or other small objects which may be in the path of the moving bat or dolphin. Edges, contours, and recessed or uneven surfaces on larger objects will be revealed more readily and more precisely by the higher-frequency sounds that they reflect; lower frequencies may give an impression of the entire object but without the details of its configuration and of its texture. The higher frequencies in the enchoranging signal are essential for fine resolution. In most microchiropteran bats and odontocetes, signal frequencies exceed 20 kHz and, in some instances, even 100 kHz. Unfortunately, high frequencies propagate over shorter distances than do low frequencies, so that, in order to be effective for guidance and prey detection, the echoranging signal must be emitted from the source at very high sound pressure levels. In bats the emitted signal in air, measured close to the animal, is often over 100 dB SPL; in dolphins sound pressures above 200 dB SPL have been recorded underwater. In spite of the high signal intensity at the source, echolocation, particularly in air, is a short-range phenomenon. The confusion caused by the echoes from more distant objects, which would interfere with the targets of concern both in and closely surrounding the animal's path of travel, is therefore reduced.

The high-frequency, high-intensity sounds used in echolocation are also remarkably brief in duration, often less than a millisecond. Brief signals are essential in order to avoid the perceptual confusion that would result from the overlap of the signal and its returning echo. As with everything else, there is an important tradeoff here, since longer signals can provide more information regarding target detail. The animals therefore adjust the duration of their echo pulse in relation to their distance from the target. The problem is exacerbated underwater because sound travels so much faster in water than in air. For example, dolphins are commonly probing their environment with acoustic pulses that travel 155 centimeters every millisecond. The signals they emit tend to be much briefer than those of bats, whose echoranging pulses are moving at a more modest 34 centimeters every millisecond. Additionally, when difficult discriminations arise or when the target distance has been considerably reduced, as in a successful chase after moving prey, the rate of emission of the pulses is considerably increased.

Echoranging signals are also highly directional. The analogy to a flashlight is a useful one. The animal is probing its environment with sound beams. Such focused sound arrives at, and is reflected from, the object of interest quickly and directly without provoking excessive interference due to echoes from the entire surrounding environment. Both dolphins and bats scan their environment so that the beam of sound they emit sweeps a wide area in the path ahead. Evidence for such precise focusing or beaming has been gathered by very careful measurement techniques together with some judicious animal training. Some interesting results are presented for a bottlenose dolphin in Figure 6.6. The animal was trained by operant conditioning techniques to place its head through a small hoop, thus keeping it in a relatively fixed position. With further training the dolphin learned to emit its echoranging pulses only on cue from the investigators. The target

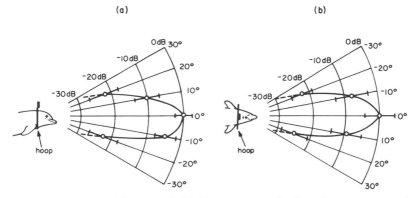

6.6 Vertical (a) and horizontal (b) beam patterns for the echoranging signal of the bottlenose dolphin. The angle in degrees of arc is indicated above and below a position straight ahead of the animal in (a) and to the right and left in (b). The dB lines or contours represent the drop (−dB) in acoustic energy with increasing distance from the center of the beam (at 0 degrees) where the energy in the echolocation signal is at a maximum. The line through the open circles represents the beam pattern and the vertical bars the variance based on a series of measurements. (After Au 1980.)

was located at a distance of 6 meters and at the same depth as the center of the hoop. The narrow, highly focused horizontal and vertical beamwidths for the echoranging signal are displayed in the figure. At a beamwidth of 10–12 degrees (± 5–6 degrees from the center of the beam in both vertical and horizontal planes) the signal energy has dropped by 3 dB; at 40 degrees it has fallen off by more than 20 dB.

We know from the results of studies of the ability of man and other animals to localize sound sources in the environment that noisy sources—that is, sounds that contain many frequencies—are more accurately localized than pure tones, or sounds containing only a narrow band of frequencies. A greater number of frequencies provides more complete information about source location. The echoranging sounds of the cetaceans that have been studied are extremely brief clicks containing a very broad band of high frequencies. Many echolocating bats employ brief pulses in which the frequency is rapidly changing or sweeping over a wide range (from 50 kHz to 25 kHz in the brown bat, for example). In both instances the spectral characteristics of the reflected sound may reveal more about the intricate details of the target than would be possible with a single frequency or even a narrow band.

As if to confound us, some bats, notably the horseshoe bats, do employ single frequencies which may be preceded and/or followed by a brief frequency sweep. Apparently the purpose of such single frequencies lies not in revealing the specific characteristics of targets but rather in informing the hunting bat about the speed of its rapidly departing prey. The principle is the Doppler effect: a constant frequency source moving relative to a receiver or listener will increase in frequency at the listener's ear if it is approaching and decrease in frequency if it is receding. The frequency transmitted from the source remains the same; it is at the observer's ear that the frequency is changing. The rate at which the frequency changes will be greater the faster the approach or withdrawal of the sound source. The bat, then, is able to use the Doppler effect or frequency shift in its returning echo to gain valuable data regarding the velocity of its moving prey. In fact, some bats are known to compensate for the frequency shift by altering the frequency of their echoranging pulse in order to maintain the echo at a nearly constant frequency as they close with their prey.

Echolocation in Bats

Many of Griffin's early experiments were concerned with confirming the hypothesis (which seemed somewhat implausible at the time) that bats, in some manner not yet understood, used their hearing in orientation and navigation. In rooms large enough that bats could fly, wires were strung from wall to wall in the manner of a simple yet effective obstacle course. Griffin was interested in the ease or difficulty with which bats in flight avoided the wires. A degree of quantification was provided by varying the size of and the distance between the wires. Subjects could be scored in terms of hits or misses and hits could be further broken down into "crashes" or merely "touches." Subjects were then either blindfolded or gagged, or their ears were plugged. As a control for the discomfort that these procedures may have caused, hollow tubes were also placed in the ears of some of the animals. Such a precaution would refute the criticism that any changes observed in the bats' behavior were the result of the discomfort imposed by earplugs.

Those animals whose eyes were covered and the controls with hollow ear tubes performed as well as the normal, unhindered animals. Animals whose ears were plugged or whose mouths were covered registered a far greater number of crashes and touches; in fact, these animals were almost incapable of avoiding the wires. Those few that did may have partially circumvented the mouth gag with sufficient sound energy to get past the ear plugs. That, however, is conjecture. In bright or dim light these same animals with earplugs and mouth gags did no better.

Donald Griffin, together with colleagues Robert Galambos and John Pierce at Harvard, had been the first to show that bats did, in fact, emit ultrasonic cries. These could be documented with the instrumentation available at that time. Using careful recording techniques, Griffin was able to demonstrate that bats with ears and mouth open were actively echolocating in avoiding the thin wires in their flight path and, further, were increasing their rate of emission of acoustic pulses as they approached and successfully avoided the wires. Animals with their ears plugged showed no such change in rate of pulse emission. The case for echolocation by bats was well established. It remained for later investigators to work out the intricate details of the bat's biosonar system.

It is difficult, if not impossible, to arrive at precise and reliable measures of the bat's acuity for the detail in the echoes it receives under natural conditions when both bat and target are moving rapidly. James Simmons at Princeton devised an ingenious yet simple arrangement in the laboratory for measuring the bat's echoranging acuity with both bat and target stationary. Although it may be argued that we cannot always generalize from the simplified "indoor experiment" to the

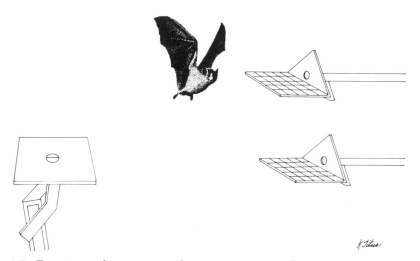

6.7 Experimental arrangement for measuring target discrimination perform-ance by echolocating bats. (After Simmons and Vernon 1971.)

complexities of the real world, the laboratory experiment does simplify conditions in order to make sense and achieve some degree of rigor, order, and control over natural phenomena. Simmons was asking a significant question about the properties of the bat's sonar receiver, and an answer was likely only under controlled laboratory conditions.

Big brown bats, blindfolded, were trained to fly from a small ele-vated platform to one of two landing platforms about 40 degrees apart on which triangular targets were mounted. The experimental arrange-ment is shown in Figure 6.7. Although the bats were required to fly to the targets, their discrimination between the targets, based on their perception of the returning echoes from their emitted pulses, was made while both bat and target were stationary. Subjects received a live mealworm for flying to the nearer of the two targets. The position (right or left) of the further target was changed on different trials, while its distance from the bat was fixed at 30 centimeters. The dis-tance of the nearer target was varied. Results are shown in Figure 6.8. When both targets were the same distance from the bat (a difference in distance of zero on the graph), the bat flew to each target in about half of the trials. At differences in distance of 5 centimeters or greater, the bat flew almost unerringly to the nearer target. Threshold—esti-mated as halfway between these two extremes, or at the 75 percent correct point—is between 1 and 2 centimeters, or about 13 millimeters. The bats performed just as well when the absolute distance of the targets was increased from 30 to 60 centimeters—an impressive demonstration of the acuity of the bat's biosonar system.

In a second, related experiment Simmons removed the targets and placed a small and very precise microphone close to the starting platform and two high-frequency speakers behind the landing platforms. With appropriate electronic circuitry he was able to delay the bat's sonar cry to one speaker or the other. The bat received its mealworm for flying to the platform and speaker with the earlier returning echo. The sonar signals transmitted by the bat were found to be similar in the two experiments. In the second experiment, the bat's threshold acuity for discriminating differences in time delay of the two signals was about 60–63 microseconds which corresponds to a difference in distance of about 12 millimeters. In other words, the echo from the target that was 12 millimeters more distant was delayed by about 60–63 microseconds relative to the echo from the nearer target. The results offer convincing evidence that bats extract information about

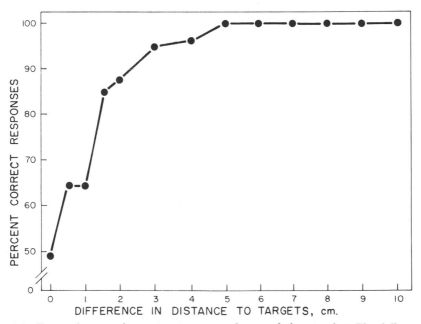

6.8 Target distance discrimination acuity for an echolocating bat. The difference threshold for distance, at the 75 percent correct value, lies between 1 and 2 centimeters at a distance of 30 centimeters from the platform. (After Simmons and Vernon 1971.)

spatial dimensions on the basis of the temporal differences in their returning echoes. The bat's auditory system must perform its analysis on these temporal differences. Further experiments by Simmons and others on the bat's echolocation system assure us that these animals are little disadvantaged without the use of sight.

Echolocation in Dolphins

Some of the earliest experiments on the use of echolocation by dolphins were carried out by Winthrop Kellogg and his colleagues at the Florida State University Marine Laboratories; these are described in his book *Porpoises and Sonar*. The dolphin story began to unfold about fifteen years after the first chapters on the use of a sonar system by bats had been written by Griffin and his coworkers. The similarities and parallels between these two apparently very different mammals are significant.

Much of the experimental work with dolphins has been carried out in large underwater pens, which may be located inside harbors or bays or may be built like large swimming pools inland, resembling those seen in commercial marineland parks. Like bats, dolphins have proved adept at avoiding obstacles without the use of their sight. The dolphins are blindfolded with large suction cups, which are easily removed after the experiment. Using such methods, Kellogg reported that dolphins have no difficulty avoiding a large array of one-inch-diameter pipes in order to obtain food or discriminating between a fish and a gelatin capsule of equal size.

In later experiments techniques were developed to keep the target and the animal stationary so that the animal's echoranging acuity could be determined more precisely. Such an arrangement is sketched in Figure 6.9. The experimental enclosure was a floating pen, in this instance located in Kaneohe Bay on the island of Oahu in Hawaii. The animal was trained to position its anterior jaw in a small chin cup which could swivel to permit acoustic scanning in the forward direction. Two underwater targets in front of the animal, positioned 40 degrees apart, were located behind an acoustic screen that was opaque to sound. On either side of the chin cup was a small paddle or response device which the animal could strike with its nose. The targets were raised by monofilament lines to the experimenter's booth. In a typical experiment, after the screen was raised, the subject indicated which target was closer by striking the paddle on the same side. Fish were given as a reward. Thresholds were determined by varying the target distance and recording the animal's response on a large number of trials.

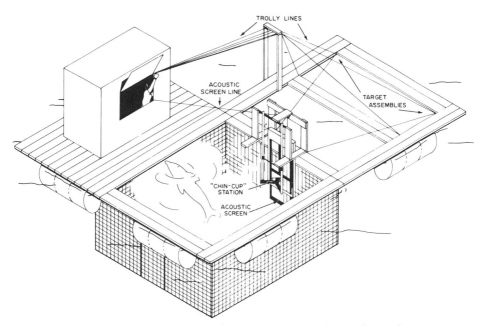

6.9   Underwater pen serving as the experimental enclosure for studying target distance discrimination performance by echolocating dolphins. (After
Murchison 1980.)

Experimental procedures such as the one just described and others
quite similar have been effectively employed in the analysis of the dolphin's acuity in perceiving the echoes from its echoranging signals.
The research until now has been much less extensive than that done
with bats. The dolphin's threshold for detection of fine wires is only
slightly greater than 1 millimeter. Russian scientists, who have shown
a considerable interest in the echoranging abilities of dolphins, estimate their useful range for echolocation to be about 5 meters for targets varying in length from 5–15 centimeters. These and other related
findings suggest (as mentioned earlier) that dolphin echolocation is
primarily effective at short to intermediate distances and is not a long-
range sensory system. The same conclusion can be drawn regarding
the sonar system of bats. However, the effectiveness and resolving
power of this acoustic guidance system at limited range, relative to
sight, are well established.
    The ability to image the environment by the use of self-generated
sound that is reflected from environmental objects represents an unusual and very special form of adaptation found in just a few species
but employed with utmost precision in only two, the bats and cetacea.

That the adaptation has been successful is clear from the striking variety of microchiropteran species in particular but also of cetaceans, and their establishment in a wide range of habitats. The success of the bats, for example, has led some scientists to suggest that their nocturnal exploitation of insect fauna in the New World was instrumental in the extinction of a sizable group of their nocturnal competitors, our own ancestors—the New World prosimian primates.

In addition to using sound in communication and in other ways common to most animals that hear, the bats and cetacea rely on sound much as the rest of us depend on light. Our movements in our environment and our detection, recognition, and discrimination of environmental features, except in darkest night, are made possible by reflected light—theirs by reflected sound. Whereas we use natural or artificial light from the sun, moon, or other source, the bats and cetacea act as both source and receiver when they bounce sound off environmental objects. Their prosperity on earth must be related to their judgment of the finest details of the structure of objects based on the echo returning from them. It is in many ways comparable to our own perceptual resolution of objects based on reflected light.

7

THE DISCRIMINATING

PRIMATES

IN CONTRAST TO the bats and dolphins with their unique acoustic adaptations, the primates are not usually considered highly specialized animals. Yet in the course of their evolution they have been subjected to certain pressures and have undergone subsequent changes which have had significant consequences for their sense of hearing. The primates descended from primitive insectivore stock some 60 to 70 million years ago near the beginning of the mammalian radiation and have developed characteristics which set them apart from other mammals.

Early in their evolution the primates moved into arboreal econiches, and thus their adaptations may to a considerable extent be viewed as successful responses to the pressures imposed by this form of environment. A large number of the early and more primitive primates were nocturnal; many of the more primitive living taxa are still creatures of the night. Other primates, including a variety of current forms, operate exclusively during the day; many of these, including the baboons, the macaques, and man, have returned to a terrestrial existence with successful adjustments to the fresh demands required by life on the ground during the daylight hours.

The living primates are grouped into two suborders—the prosimians and the anthropoids. The prosimians, sometimes called half-monkeys, are the pottos and bushbabies from Africa and the slow and slender lorises from Asia; the tarsiers from Asia and the Philippines; and the lemurs, which include an amazing variety of species all living off the eastern coast of Africa on the island of Madagascar (the Malagasy Republic). The anthropoids comprise not just the African and Asian apes and man, but also the New World monkeys of Central and South America and the Old World monkeys of Africa and Asia.

If the prosimians are considered the more primitive of the primates resembling ancestral forms, then perhaps we can see, when comparing them with the monkeys and apes, some of the trends that have occurred during primate evolution. Most of the prosimians have maintained a nocturnal existence, while only one species of anthropoid (the owl monkey of South America) has done so. Significant changes in the sensory apparatus, in the ways in which the primates received signals from their environment, took place in the course of their evolution. Seeing became increasingly important for the primates because of the demands of a precarious life high above the ground and, too, because of their movement into diurnal (daytime) adaptive zones.

Enhanced visual acuity and color vision, not highly developed in most mammals, appeared in the higher primates. As the orbits rotated more toward the front of the head (compare the anthropoids with the prosimians) the visual fields of each eye overlapped, permitting stereoscopic vision and therefore better depth perception—particularly useful in high places. The sense of smell, so critical in the terrestrial mammals, apparently became less important and less relied upon by the primates; the area of the brain devoted to smell became proportionately smaller. At the same time, those portions of the brain related to vision became relatively larger. The changes that occurred in hearing, as we shall see, were more subtle. In the anthropoid primates, at least, these changes may have been related to the pressures for a more intricate intraspecific communication system. These, in turn, were a consequence of the demand for a more cohesive social organization in those primates that spent more time on the ground where predation increased and food sources were often more widely distributed.

Primates evolved digits with nails in place of claws, a development that is related to their considerable manual dexterity. A trend toward upright posture and locomotion and an increase in brain size, in addition to an extended period of infant development under the close scrutiny of mother, siblings, and relatives, are characteristics which serve to distinguish the primates from other mammals. With the modifications that have occurred in their hearing and in their communicative skills, it is easy to see the importance of characteristics, such as brain development and a prolonged childhood, which could be expected to facilitate learning.

AUDITORY STRUCTURES IN PRIMATES

In its gross features, and in many of its microscopic ones as well, there is little to distinguish the primate ear from that of other mammals (see Figure 5.1). Sound waves are funneled into the ear canal past the pinna, which appears to be less mobile and therefore perhaps less helpful in directional hearing in the primates—at least in the anthropoids—than it is in many other mammals. In the ear canal, certain acoustic frequencies (in the region of 2 kHz for man) are amplified as the ear drum receives the pressure wave and transmits it via the three-bone ossicular chain to the oval window of the inner ear, where further amplification takes place due both to the areal ratio between ear drum and oval window and to the lever action of the ossicles. In the fluid-filled cochlea of the inner ear (see Figure 5.2), the hair-bearing ends of the receptor cells on the basilar membrane are displaced or mechanically deformed as a consequence of the sound-induced traveling wave which engenders a relative motion between the hairs of the sensory cells and the overlying tectorial membrane. The end result is stimulation of the individual auditory nerve fibers that depart from the base of the sensory cells and subsequently transmit the information to the nervous system and brain.

Physiological function and the encoding of the various parameters of the acoustic signal (frequency, intensity, and so on), are much the same in the primate ear as in the ears of other mammals. The unique specializations that we find in the kangaroo rats, the bats, and the dolphins are not readily apparent in the primate ear. The subtle differences that have been discovered between the hearing acuity of primates and that of other mammals are perhaps more likely a function of the differences in the nervous system and brain, where the integration of more complex signals occurs. The processing of the intricate and biologically relevant components of human language provides one example.

HEARING IN PRIMATES

Primates, *Homo sapiens* included, have proven extremely adaptable not only in their interaction with the natural world but in their dealings with laboratory scientists who have asked them an endless variety of questions about their sensory experiences. Through the methods of animal psychophysics (see Chapter 5), we have obtained a relatively clear picture of the acoustic capabilities of several primate species and

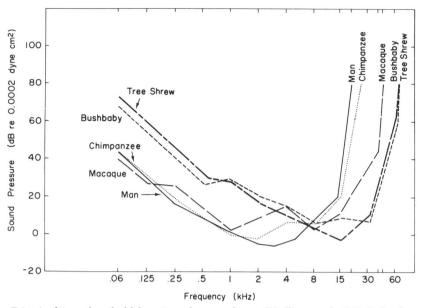

7.1 Auditory threshold functions for tree shrew (Heffner et al. 1969a), bushbaby (Heffner et al. 1969b), macaque (Stebbins et al. 1966), chimpanzee
(Elder 1934; Farrer and Prim 1965), and man (Sivian and White 1933). The
tree shrew is an insectivore, the bushbaby a prosimian primate, and the macaque monkey, chimpanzee, and man are anthropoid primates. (After Stebbins 1971.)

at least preliminary accounts of many others, including the prosimians. Measures of threshold sensitivity and frequency range taken
from several very different species are shown in Figure 7.1; these
measures are representative of both prosimians and anthropoids. With
their extended high-frequency hearing (60–70 kHz), the prosimians
resemble other nonprimate mammals such as the insectivorous tree
shrews (*Tupaia*), which are included here for comparison. In sharp
contrast, our own hearing extends to only about 20 kHz. In between
are the chimpanzees with an upper limit of 30–35 kHz and the Old
and New World monkeys with hearing to 40–45 kHz. It is tempting to
suggest an evolutionary progression, even though it is in a retrograde
direction. We will return to this unusual development later. Considerable sensitivity at lower frequencies, in the range of their communication sounds, is characteristic of at least the anthropoid primates; in
this respect, they do not differ greatly from some of the other terrestrial mammals such as the cats.

In many of the laboratory experiments from which these observations come, monkeys are conditioned by methods similar to those applied to guinea pigs (Chapter 5). There are a few important differences. The monkeys are trained by positive reinforcement or reward procedures to enter a special restraining chair for daily training or testing. Standard earphones may then be placed directly on the monkeys' ears, so that each ear may be tested separately. A flashing light in a hollow tube or cylinder in front of the animal is the cue to touch or grasp the outside of the cylinder. The light then stops flashing and becomes steady, providing the animal with the information that it has made the correct response. If the monkey continues to hold on to the cylinder, a brief acoustic signal is presented. Now the animal releases its grip on the cylinder and is immediately reinforced with food. Should the monkey let go prematurely or too late, after the signal has been turned off, a punishment in the form of a 6–8-second time-out follows. Like the guinea pig, the monkey learns quickly and, in time, becomes a reliable and accurate observer of acoustic events, so that many facets of its hearing may be examined under controlled laboratory conditions.

Following training, thresholds are determined in an interesting manner by establishing a feedback loop between subject and stimulus. The testing process begins with the stimulus presented at a moderately high sound level. If the monkey responds correctly, it is reinforced with food and the stimulus intensity is lowered by a fixed amount on the next trial. The procedure continues until the animal fails to let go of the cylinder, indicating that it has not heard the signal. On the following trial the stimulus intensity is raised. In this way the sound level changes according to the animal's response, and yet his response is surely determined by the intensity of the stimulus. An example of the stimulus progression is shown in Figure 7.2. Threshold is calculated as the average of the transitions (indicated by the arrows in the figure) between "heard" and "not heard," and it is easy to see why the procedure is often called the staircase method (also known as the tracking or titration method). It is one of the quickest and most efficient ways of determining threshold, yet it is difficult for the subject because the stimulus is almost always near the threshold and therefore difficult to detect.

Knowing an animal's absolute sensitivity and frequency range tells us very little about its capability for fine discrimination of acoustic frequency or intensity within that range. The hearing of the noctuid moth, for example, covers a frequency range far more extensive than many vertebrates, and yet its powers of discrimination for frequency

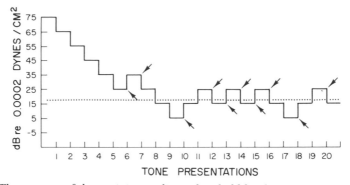

7.2   The process of determining auditory threshold by the staircase or tracking method. Correct detections cause the tone to be attenuated, whereas failure to hear is followed by an increase in tone intensity on the next trial.
Threshold is indicated by the horizontal dotted line. The arrows indicate the
transition points between correct detections and failures to hear. (After Stebbins 1973.)

or intensity are almost completely lacking. Two tones widely separated
in frequency sound very similar to a moth. However, when the bat's
frequency-modulated hunting cry sweeps the frequency range to
which the moth is sensitive, the moth is quick to respond (see Chapter 2). In the moth's environment, the selective pressures to evolve a
more finely tuned or discriminating auditory system have apparently
not been especially intense.

In the primates, however, particularly in the more terrestrial anthropoid primates (Old World monkeys and apes), the requirements
of life on the ground and the attendant social organization have been
followed by a variety of successful adaptations, not the least of which is
an elaborate system of intraspecific communication. It is communication that renders the social group effective and cohesive. Further, the
level or complexity of the group's organization is directly related to the
complexity of its communication system, which may be defined with
reference to the nature and extent of its repertoire of sensory signals.
More elaborate communication systems make greater demands on the
discriminative acuity of the sensory systems.

In discussing the evolution of hearing and acoustic communication, we are confronted with the question of chicken versus egg. Did
the selective pressures on social organization and communication in
the primates provide an advantage for the more acoustically acute, or
did better hearing precede developments in the vocal mechanisms
underlying the production of more complex and diversified communi-

cation signals? Some scientists shudder at such arguments because they—that is, the arguments—are somewhat simplistic and can never be resolved with any degree of finality. The issue is raised here more to highlight the continuing interaction, in the course of evolution, between hearing and acoustic communication rather than to suggest that one led the other. It was an interaction that, in time, produced changes in both hearing and acoustic communication.

An acoustic communication system, rich in the variety and diversity of its signals, will likely require more of the discriminative capabilities of its listeners than one that is relatively impoverished and that employs a limited number of discrete signals. At one extreme is human language, with its vast catalog of significant sounds often with only the most subtle of acoustic differences between them. Though other primates may possess simpler systems, many employ numerous calls or sounds that are graded and that frequently differ only slightly along many acoustic dimensions (frequency, intensity, spectral composition, duration, rise time, and so on). What we might look for, then, in the primates is evidence of their capability for discriminating minute differences in the acoustic properties of communication signals.

Discrimination or difference thresholds for sound frequency, intensity, and other acoustic properties may be determined in nonhuman primates with only a slight variation in the procedure for measuring sensitivity described above. In measuring frequency difference thresholds, a repeating, pulsing tone (the standard) is presented. Shortly thereafter, a second tone, the comparison, differing in frequency from the first, alternates with the standard twice in the same repeating temporal pattern as the standard. If the monkey responds to the change in frequency, it is reinforced; if not, the second tone is turned off and the standard is again presented alone. In determining the difference threshold for frequency, each correct response to the change in frequency has the effect of diminishing the difference between the standard and comparison tone on the following trial, whereas the difference is increased if the monkey fails to respond to the frequency change. The difference threshold is the smallest difference in frequency to which the animal is able to respond correctly 50 percent of the time.

Figure 7.3 shows difference thresholds at various frequencies of the standard stimulus for Old World monkeys and man. In both, discrimination is most acute at the lower frequencies in the range of the species communication sounds and becomes increasingly poorer at the higher frequencies. Another conventional manner of examining these data is to consider the frequency difference threshold (Δf) as a

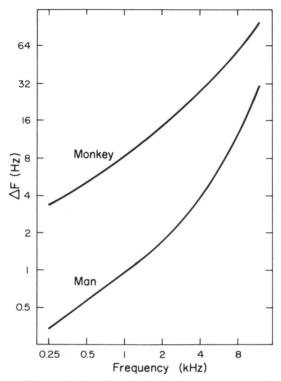

7.3   Frequency discrimination thresholds for the macaque monkey and man. Range is between 250 Hz and 8 kHz. (After Nordmark 1968; Stebbins 1973.)

fraction of the standard frequency ($\Delta f/f$). When the latter quantity is observed across the audible frequency range, a decrease in acuity is still seen at the higher frequencies, although the effect is somewhat reduced. If we now compare a prototypical terrestrial mammal, such as the tree shrew, with three very different species of primate—bushbaby or galago, Old World monkey, and man—we find clear differences in discrimination (see Figure 7.4). It can be argued plausibly that the ability to resolve very small differences in acoustic frequency is related to the level of complexity of an animal's communication system. Despite the compelling aspect of the evidence, it is circumstantial and the argument is speculative. But what else would require humans to discriminate a difference of only 2 Hz at 1,000 Hz?

A related puzzle is why, in the course of evolution, high-frequency sensitivity decreased in man and the other anthropoid primates (see Figure 7.1), since it seems to be correlated with an enhanced differential acuity for frequency. It is as if the putative "higher primates" gave up high-frequency hearing for improved frequency discrimination within a more restricted, lower-frequency range. The structural and physiological changes in the inner ear and auditory nervous system that subserved these changes in hearing are unknown. If, as in Chapter 5, it is argued that the earliest mammals evolved high-frequency hearing so that their small heads cast a significant sound shadow, providing a discriminable intensity difference between the two ears for accurate sound localization, then perhaps the apes and monkeys, with their larger heads, no longer required the high-frequency sensitivity of their smaller ancestors.

The ability of the primates to resolve very small differences in acoustic frequency, in intensity (differential intensity thresholds for monkeys are on the order of 2 dB, and for humans less than 1 dB

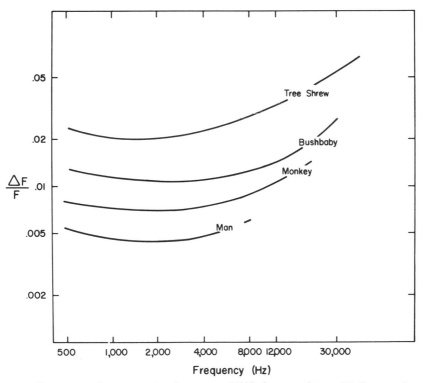

7.4 Frequency discrimination functions ($\Delta f/f$) for tree shrew (Heffner et al. 1969a), bushbaby (Heffner et al. 1969b), macaque (Stebbins 1978), and man (Filling 1958). (After Stebbins 1978.)

across most of our audible frequency range), and in the fine temporal structure of auditory signals may facilitate the perception of complex communication signals. But such capability may also function importantly in sound localization, where the cues that reveal the location of a sound source are the temporal separation between the two ears (sound arrives later at the more distant ear) and the spectral and intensive differences between the ears produced by the head acting as a sound shadow. Consequently, the acuity and accuracy of an animal in localizing a sound source in three-dimensional space are directly related to its ability to discriminate these three parameters of an acoustic signal—frequency, intensity, and time or temporal disparity between the two ears.

Although sound localization is an important function in the natural world, it can be evaluated precisely only in a controlled laboratory environment where other potential cues can be eliminated or their contribution measured. Ordinary walls reflect sound, and these reflected sound waves or echoes make accurate localization of an acoustic signal in a small enclosed room difficult. The acoustic conditions bear little resemblance to the outside world of grass, trees, and sky. Fortunately it is possible to construct a highly specialized room where most echoes are excluded. An anechoic room is lined with specially designed fiberglass wedges on all sides, ceiling and floor included; the floor is usually covered as well with a metal grid or soft rug. Most of the emitted sound is absorbed so that when the signal is presented it will not be confused with its echo at the listener's ear. Although natural conditions are more closely approximated in such a room, this is not the real world with its trees, bushes, and other textured and shaped reflective surfaces. Like most laboratory situations, use of the anechoic room represents an attempt at analysis of the complex phenomenon of acoustic localization under the "purest" conditions possible. It is, at best, a compromise.

Measurement of localization acuity presents a spatial discrimination problem. The smallest discriminable difference between two discrete sound sources located in different positions in the animal's horizontal or vertical plane can be evaluated by determining their angular separation in degrees of arc. The sources or speakers are equally distant from the subject. The distance between them is calculated on the basis of the angle between an imaginary line from each speaker to the center of the subject's head. The example shown in Figure 7.5 involves horizontal localization. Usually one speaker serves as a standard or reference and is positioned directly in the subject's line of regard or at an azimuth of zero degrees. The second speaker is then varied in its angular distance from the standard and the subject's re-

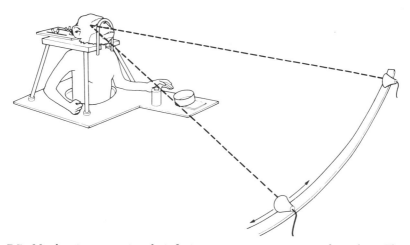

7.5 Monkey in a restraint chair facing an arc containing sound speakers. The speaker to the reader's right is in the subject's line of regard and therefore at zero degrees azimuth.

sponses noted. Accuracy improves with increasing speaker separation. In one procedure, the subject is simply required to respond when the sound shifts from its central position directly ahead to another location. The threshold of localization, often called the minimum audible angle (MAA), is the angular separation between the two speakers that the subject correctly reports in 50 percent of the trials. Variation in speaker separation yields a familiar psychometric function (see Figure 7.6). In this instance the subject was a macaque monkey responding to the change in the horizontal location of a band of noise. The threshold determined from the function was about 12 degrees. Of course, localization threshold will vary depending on the nature of the signal, in addition to many other factors.

Once again, conditioning procedures permit us to question animals directly about their sensory capabilities. Figure 7.7 shows the set up for a sound localization experiment, in which a monkey is placed in a special restraint chair in an anechoic room. The restraint is necessary in order that the animal's head be fixed in position and directed at the speaker straight ahead (at zero degrees azimuth). Monkeys are trained with food reinforcement to enter these chairs—a task which they master quickly. They then learn to place their hand on a small disk (see Figure 7.7) in response to a light that appears under the standard speaker directly ahead. A pulsed sound is immediately presented from the speaker. If the subject keeps its hand firmly in place on the disk, the sound will eventually shift to one of several other

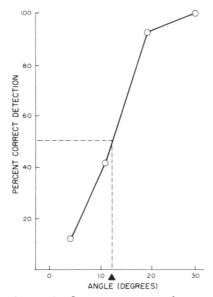

7.6 Sound localization acuity for a macaque monkey, measured by angular distance in degrees of arc from a reference speaker at zero degrees azimuth and determined from this psychometric function. Threshold is indicated by a dashed line and arrow. (After Brown et al. 1980.)

speakers located on an arc to the animal's right. Each speaker is situated at a given angle relative to the first or standard speaker. If the monkey lifts its hand when the source of the sound shifts to another speaker location, it is reinforced with food. Should the monkey fail to respond by lifting its hand, the sound source returns to the standard or reference location without reinforcement. At wide angles, animals have little difficulty in reporting the change in location; as the angular separation between speakers is decreased, the animals find the task increasingly difficult and their performance deteriorates accordingly. Over a series of trials, sound source position is varied, the subject's responses are recorded, and threshold is determined from a psychometric function similar to the one shown in Figure 7.6.

Of what actual importance is the ability to localize sound to the behavioral biology of the animal? Zoologist Peter Marler has emphasized the three-way relationship between localization, the acoustic structure of a signal, and the function of that signal for species members. Some sounds, depending on their physical properties—spectral composition, intensity, and so on—should be more accurately localized than others. At the same time, it would be distinctly advantageous for many animals to be able to make some calls that are easily locatable but others whose location is difficult to determine. Certain alarm calls, for example, may be acoustically designed to be cryptic in order not to reveal the location of the caller, who is naturally at some risk from predation. On the other hand, the success of calls such as those that function in mating and reproduction or other close forms of social interaction is largely dependent upon the ease and accuracy with which they can be localized.

The complexity of such problems underscores the diversity of strategies and approaches necessary for their solution. Both laboratory and field methods are essential. Cooperation between investigators in these two areas can only enhance our understanding of the biological significance of sound localization. For example, the relationship between signal structure and localization acuity or accuracy was suggested on the basis of extensive field observations; yet this is a question that requires the analytic and control techniques of the laboratory for its answer. On the other hand, the matter of signal function is one that for the most part must be settled by the observational methods of field study.

From the field, Marler has suggested the features of a call that

7.7 Macaque monkey seated in a restraint chair in an anechoic room.

might render it more difficult to localize. Early laboratory research in which human subjects were required to localize tonal stimuli had indicated the importance of two significant cues for sound localization—time differences for the signal reaching the two ears (Δt), a most effective cue at low sound frequencies with their relatively long wavelengths, and intensity and frequency differences at the two ears (Δfi), a salient cue at high frequencies (short wavelengths) because of the sound shadow cast by the head. On the basis of these earlier findings, we might predict that relatively noisy sounds encompassing both high and low frequencies, and therefore making both Δt and Δfi cues readily available, would be easily localized. In contrast, calls or sounds that are more tonal in nature and that possess a limited spectral range would be considerably more difficult to localize, particularly if such sounds were composed of middle-range frequencies between the low frequencies where the Δt cue is operable and the high frequencies where the Δfi cue is functioning.

The hypothesis of the relationship between acoustic structure (in this instance the frequency bandwidth of the calls) and localization acuity was put to the test at the Kresge Hearing Research Institute at the University of Michigan. The auditory signals were macaque clear or "coo" calls often used to maintain contact or to solicit attention within a monkey troop. Some calls were tonal in nature, containing few frequencies; others swept over a range of frequencies, thus providing considerable spectral content. Macaque monkeys served as subjects in an anechoic room under the experimental arrangements described previously. Some of the findings of these experiments are presented in Figure 7.8. Psychometric functions are shown beside the sonographs of the calls used as signals to be localized; thresholds, determined directly from the functions, are indicated by dashed lines and arrows. Minimum audible angles (thresholds) for the more tonal signal (narrower bandwidth) are about 15 degrees of arc; for the frequency-modulated signal it is 4 degrees. The broader bandwidth call is localized more accurately, thus confirming the predictions based both on field work and on earlier experiments on localization with human subjects.

COMMUNICATION IN PRIMATES

Of all animals, primates are probably the most gregarious. Their communication systems appear to be more complex than those of other mammals and to be intimately related to their well organized and

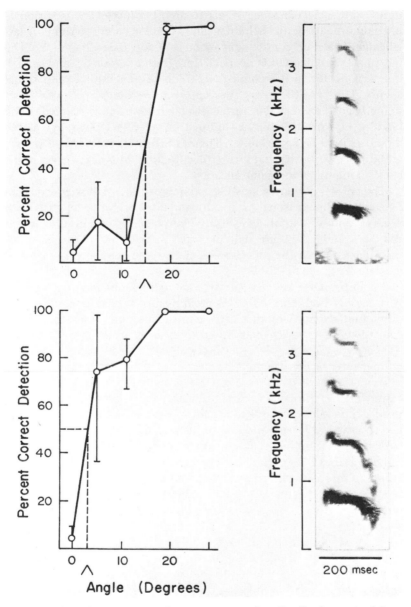

7.8 Sound localization acuity for two macaque "coo" calls, determined from psychometric functions (on the left). Threshold (MAA) in degrees of arc is indicated by dashed lines and arrows. Sonographs of the calls are to the right of the appropriate psychometric function. Points are average values for three animals; vertical bars represent the range of values. (After Brown et al. 1978.)

tightly knit social existence. In discussing their use of acoustic signals
in communication, it is helpful again to employ Peters' system of classification based on context and signal function (see Chapter 5). The
message systems include neonatal, integrative, agonistic, and sexual;
the terms refer to the context in which the communication signal
occurs. The vervet monkey (*Cercopithecus aethiops*) provides a particularly good example because it has been so carefully studied in the
field by the Animal Behavior Group at Rockefeller University (whose
members include Peter Marler, Thomas Struhsaker, Dorothy Cheney,
and Robert Seyfarth) and further documented by Roger Peters in his
book on mammalian communication.

The vervet, also known as the guenon or green monkey, is an apt
choice for other reasons. As an African Old World monkey, it is both
arboreal and terrestrial; it is versatile and facile on the ground and in
the trees. Indeed, its life style may well parallel that of our own primate ancestors in the process of adapting to a completely terrestrial
existence.

Infant primates are characterized by an extended nurturant period
with mother and other kin. This prolonged relationship, compared to
other animals, may enhance early learning and may at least partly account for the extensive communicative repertoire of primates. Even
the very young make use of a variety of calls in solicitation of contact
or in distress. The cooing and babbling of human infants is well
known. Tonal coo calls are uttered in a similar context by many macaque infants; the same general call with varying acoustic structure is
used by juveniles and adults for many different functions. Struhsaker
has recorded at least five different distress calls by infant vervets related to mother-infant separation. As the distance between mother and
infant increases the "rrah" call changes to "eee" or "rrr" with an increase in intensity. Alternation of these latter two calls provides a signal that the mother should have no difficulty localizing, with its combi-
nation of high and low frequencies. Should the separation continue,
the infant's distress call shows further shifts in acoustic structure
with greater increases in intensity and in the frequency of the call. A
high-pitched squeal or scream that may follow forced weaning represents the ultimate in infant vervet distress. A somewhat guttural or
throaty "eh-eh" greets the vervet mother upon her return.

Integrative message systems in primates are rich and diversified in
content. Perhaps the most intensely and thoroughly documented have
been the contact solicitation or coo calls of macaques and the alarm
calls of vervets, both studied by members of the Rockefeller group.
Steven Green's extensive field work with the Japanese macaque has

given us a key to understanding the graded nature of many animal communication systems. Again the acoustic structure of the call, here the coo, apparently plays a major part in determining its function. An example or variant of such a call is shown in sonographic form in Figure 7.9. Similar to the calls shown in Figure 7.8, its detail is enhanced. The darker bands signify those frequency ranges that contain the most acoustic energy. Note that during the call the frequency increases and then returns to its original level, and that this change is particularly noticeable in the second band (counting from the bottom). Relative to the duration of the call, the frequency change is gradual in both directions.

On the basis of his field observations of Japanese monkeys, Green has suggested that the temporal position of the frequency inflection, or peak, is a key to the content of the message that is being transmitted. Calls with an early peak serve an integrative or cohesive function for the group. Individuals or subgroups separated from the troop on the move continue to maintain contact by use of these calls, which are often answered by other animals in the larger troop. When the frequency modulation occurs late in the call, a higher state of arousal is often indicated and the call may serve an agonistic or sexual function. A subordinate animal may give this variant of the coo call in response to the approach or presence of a more dominant animal, as if to appease and to prevent a fight. It is also a call used by females in estrus in solicitation of a qualified male. Thus, an apparent grading in one acoustic feature, the change in position of the frequency peak, is sufficient to produce major changes in signal function. The parallels to our own communication system are striking.

One of the more exciting recent findings is the vervet monkey's use of different alarm calls for different predators. If, as Seyfarth and Cheney have suggested, this can be considered an instance of signs referring to objects in the external world, then perhaps we have an example of semantic communication in animals other than man. The different alarm calls are clearly discrete and apparently not part of a graded system, as are the coo calls. Initial field observations revealed that the vervets sounded the alarm with a distinct cry for each of three different predators—mammalian carnivores (leopards, lions, and so forth), raptors such as eagles, and snakes. Occasionally, alarm calls were heard in the presence of humans or baboons, but the instances were too few for reliable data collection.

The alarm call that warns of a terrestrial predator resembles a bark, or, if the caller is a female or young male, it may sound more like a chirp. As Struhsaker first described it, the call given by an adult male

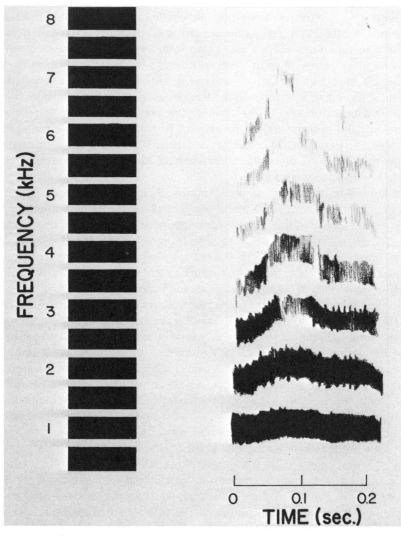

7.9 Sound spectrograph of Japanese monkey coo call. The dark bands indicate the frequency regions containing the most acoustic energy. (After Green 1975.)

is relatively long and tonal and is punctuated by a rapid sequence of inhalations and exhalations, ranging in frequency between 500 and 1,500 Hz. The snake-alarm call consists of a series of much briefer sound pulses covering a much wider frequency range (often extending above 16 kHz) and has been described as a "chutter." In contrast, the eagle-alarm call is of somewhat lower frequency (200–1,000 Hz) with brief acoustic pulses or grunts combined in a short phrase ("rraup"). The calls were easily distinguished by human observers, and, based on what is known about vervet hearing, monkeys should have no difficulty telling the calls apart.

The responses of the other monkeys to the alarm calls suggest the semantic nature of this particular form of communication. When the ground-predator alarm call was given (most commonly for a leopard), the monkeys, if they were on the ground, climbed the nearest tree. If already in a tree, they either failed to respond to the call or climbed higher. The most frequently observed response to an eagle-alarm call was looking up and/or seeking cover in a nearby bush. Following a snake "chutter" the monkeys looked down toward the ground, often approaching the snake and, in a group, actually mobbing it.

Even the most careful field observations failed to guarantee a precise relationship between the alarm call and the subsequent response of the animals that heard it. Other contextual features may have been at least partly responsible. For example, animals other than the caller may have spotted the predator. In order to reduce or eliminate the effects of some of these contextual variables, Seyfarth and Cheney devised a series of playback experiments. By recording and then later playing back the alarm call in the absence of any real predator, visual and other cues associated with the predator were effectively removed; the alarm call as a communicative signal stood by itself. Upon hearing the recorded calls, the animals looked toward the loudspeaker and around their environment and then responded in the same manner as they had to the live alarm call. For the ground-predator alarm call they took to the trees, for the eagle-alarm call they sought cover in dense foliage, and for the snake "chutter" they looked toward the ground. The evidence for a form of semantic communication is convincing. Vervets appear to classify alarm calls purely on the basis of their acoustic structure even when the call itself is removed from its visual context.

Agonistic messages in monkeys are fairly common and often combine signaling in a visual as well as an acoustic mode. Such messages often function in territorial and other encounters between troops, as well as within a single troop in the maintenance and establishment of dominance hierarchies and in conflicts involving juveniles and older females. Many offensive threats are visual rather than auditory and include such gestures as movement by a dominant animal toward another animal lower in the dominance hierarchy, or an open mouth with canines exposed and head thrown back. In vervets, and to some extent in macaques, acoustic signaling is used more often by a submissive animal as a defensive threat. Vervets make use of several such calls, which include a low-frequency, almost staccato-type noise (frequencies below 1,000 Hz), a higher-frequency squeal (about 5,000 Hz), and a longer, more strident scream similar to that used by infant animals in distress. These calls may be combined in various arrangements. Generally, the gravity of the situation is reflected in the intensity and frequency content of the call, much as it is in our own communication system.

Although physical aggression between troops is even less frequent than within a single troop, there is a considerable exchange of threats between the members of two different groups that come into contact. The function of these interactions seems quite clearly to be territorial. Females and young males are major participants in these encounters and are often the first to report the appearance of another troop in the area; for this, they use a relatively intense, low-frequency, noisy "aaar" call. When the two troops actually meet, there is a very active and noisy interchange with loud barking, in addition to grunts and "chutters" that are heard only in such intergroup meetings. Other calls, similar in acoustic structure to alarm calls, may function not only as threats but as appeals for help from group members. These interactions between groups continue until one troop, often the one consisting of the transgressors, retires from the field.

In vervets and in many other monkeys, sexual messages are most often conveyed by visual and olfactory signals. In vervets one of the few acoustic signals related to sexual behavior is the scream or shriek uttered by an unwilling female in response to persistent advances by a male. It is similar to a defensive threat and therefore conveys more of an agonistic than a sexual message. It seldom fails to drive the male away.

Laboratory studies of hearing, described earlier in this chapter, generally support the findings and implications of the field research on primate communication. Primates have the basic acoustic capabilities to detect and to discriminate between the various communication calls to which field workers have assigned different message functions. Yet the findings of most field studies are based on observation and are most often correlational—that is, a given call type is associated with a given function or response on the part of an animal other than the caller. Other contextual influences (visual, olfactory) are always at work, and it is difficult to separate these from the effects of the acoustic signal per se. Can these complex acoustic signals, very different from the simple pure tones used in the laboratory, stand alone? Can they be separated from the context in which they occur? The results of the playback experiments of Seyfarth and Cheney suggest that they can. It is at this point that laboratory research can make a significant contribution. Green's field work on contact or coo calls in Japanese monkeys indicates that slight variation in the acoustic structure of the call can change the message content considerably. Is the message conveyed only by acoustic signaling or do other nonauditory stimuli play a significant role? It becomes important to know whether animals, in fact, are able to make some of these more subtle discriminations; this is a problem that can be directed to the laboratory. The complex acoustic signals themselves must be used in place of simpler stimuli such as tones, but the testing procedures are the same as those used in determining an animal's basic acuity or threshold sensitivity.

In collaboration with the Animal Behavior Group at Rockefeller University, our Primate Laboratory at the University of Michigan designed experiments to examine the Japanese monkey's ability to discriminate between different variants of the coo call in the laboratory. On the basis of Green's functional classification of these calls in the field, we selected, as stimuli, calls with an early peak or pitch shift (early high) to be discriminated from those with the peak occurring late in the call (late high). As in the procedures for determining difference thresholds for pure tones, the monkeys were trained to respond when an early high was presented after a series of late highs or when a late high followed several early highs. Many examples of each type of call were used. As controls, other macaques (not Japanese, or *M. fuscata*) served as subjects in order to answer the question of whether these calls had a special communicative significance for Japanese monkeys. The results clearly indicated that they did.

The Japanese monkeys learned the discrimination with great alacrity and had no difficulty in discriminating when numerous examples of early and late high calls were added. The other macaque species had great difficulty in discriminating between the different calls and were able to do so with some success only after extensive training over several months. The key acoustic feature presumed responsible was simply the temporal location of the peak or pitch shift within the call. Shifting this feature had communicative significance for the Japanese monkeys but not for the other species. We then changed the rules of the game with fresh subjects. In this second experiment, the animals were trained to discriminate not on the basis of peak position but, instead, with regard to the starting frequency of the call. The results were quite different. The non-Japanese monkeys easily learned the new discrimination, based on pitch. The Japanese monkeys, on the other hand, apparently found the task much more difficult to learn and made many errors in the process.

The results of these experiments are suggestive, if not conclusive. Clearly, Japanese monkeys are able to discriminate between the two call types that contain different messages based on the field classification system of Green. Furthermore, they are able to do so when the calls are taken out of their natural context, which suggests that the acoustic cues can stand alone and do not require support from visual or olfactory stimuli. This is not to say that these nonauditory cues provide no supportive function but that they may be redundant. Finally, based on the control subjects (the non-Japanese monkeys), there does seem to be perceptual specialization for these particular calls based on peak position. Whether the ability to distinguish this particular feature is learned or innate is unknown.

One of the most intriguing findings from these experiments was that the Japanese monkeys performed better when the stimuli were presented to their right ear as opposed to their left. Right-ear advantage strongly implicates the left hemisphere of the brain as instrumental or dominant in the processing of these species calls. The control subjects showed no such ear advantage. There is a striking parallel between these results and those reported in numerous experiments with human subjects, demonstrating the specialization of the brain's left hemisphere for human language. This is the first bit of evidence in nonhuman primates for cerebral dominance in the perception of communication signals. It is, of course, only a beginning, but perhaps it begins to erode the anthropocentric view that human language is what distinguishes man from beast. Language per se may be a species-specific trait, but there is little to suggest that it signifies a qualitative difference between humans and other animals.

Human primates may appear to have received short shrift in this chapter. Human hearing, speech, and language are amply treated in many excellent books and scientific articles (see the list of suggested readings). Human hearing and the human auditory system differ in subtle ways from those of the other primates. The structure of acoustic communication signals is vastly more complex and the diversity considerably more extensive in humans than in other animals. Then, too, the variety in the content of humans' intraspecific messages is several orders of magnitude greater than anything found in the communication systems of other animals. This does not provide grounds for revoking the Darwinian model of species continuity, but it considerably enhances the challenge in the search for a more coherent understanding of the evolution of human hearing, language, and thought.

8

EPILOGUE

No one has yet published a book on the comparative biology of hearing. The closest thing to it is the excellent *Waves and the Ear,* written more than twenty years ago by Willem Van Bergeijk, John Pierce, and Edward David. But their book differs from this one in at least two significant respects: on the one hand, it is oriented toward human hearing and is less comparative in its approach; on the other, it is strongly mechanistic, in that it emphasizes the important search for the physiological and morphological basis of hearing. Interestingly, it is this attention to underlying mechanisms—particularly, though not exclusively, in the auditory periphery—which dominated hearing research in the early 1980s. Perhaps this is as it should be, but while paying physiological and morphological reductionism its due, I have tried in this book to underscore the importance of the biological significance of hearing and communication as effective forms of adaptation to the environment, and to put forth the behavior of the intact animal either in the laboratory or in nature as a subject not only worthy of study but needing attention. After all, it is the behaviors we call hearing and communicating to which all discoveries of underlying mechanisms must ultimately refer.

Peripheral auditory systems are mechanical devices constructed in a finite number of different designs as pressure and displacement sensors. Outwardly simple, their complexity is due in part to their analytic ability in taking apart the various dimensions of the incoming waveform and performing an energy transformation for subsequent processing by the central nervous system and brain. There the information is digested and may then be deployed to the muscles with the appropriate command to flee, chase, seize, wait, call. Although we may never know with certainty, it seems plausible that auditory receptors evolved from more simply constructed, less sensitive detectors of substrate vibration or change in body position. Hearing as a response to changes in sound pressure is apparently not universal among invertebrates, but it is ubiquitous among living animals with backbones. It may be a Johnny-come-lately among the senses, but all of the more highly advanced forms have it and use it to good advantage.

The evolution of the vertebrate inner ear is still a well kept secret. The suggestion that it is the outcome of a gradual enfolding of the "more primitive" lateral line organ is appealing, but the evidence is not sufficiently compelling. The homology between inner ear and lateral line is not firmly established. What seems somewhat more certain is that the auditory part of the inner ear developed in the course of vertebrate evolution as an expansion of the vestibular or balance system—semicircular canals, utricle, and saccule (see Chapter 3). The origin of the vertebrate middle ear with its recorded legacy in bone is more apparent. The sequence leading from the gill slits of primitive marine fish to posterior jaw support systems to middle ear ossicles in land vertebrates, when more efficient jaw articulation rendered such posterior support obsolescent, is one that most researchers would accept.

It is a speculative though not unreasonable assumption that the earliest ears were instrumental in the detection of prey, the evasion of predators, and the acquisition of a mate. Throughout evolution, accurate localization of these sound sources required a more sophisticated, more discriminating auditory system. Simple detection and even recognition were insufficient. Animals sensitive to small differences in the different parameters of acoustic signals (frequency, intensity, and so on) had a clear adaptive advantage.

Perhaps with these improvements in discriminative acuity related to sound localization came the opportunity for animals to employ acoustic signaling in a different context as both sound source and receiver. These events awaited the development of an adequate vocal motor control system that exploited the outflow from the respiratory system as a source of acoustic energy. Together the ability to transmit an extensive variety of acoustic signals and the perceptual endowment to receive and to discriminate between them formed the basis of a viable communication system. Yet the evolution of such a system is predicated on the advantages that are conferred on its users. Mutual benefits must accrue to both signaler and receiver; the change in the behavior of the receiver brought about by the message must be profitable to both.

The success of many birds and primates has been due in no small measure to the richness of their vocal repertoire and to their considerable perceptual abilities in the discrimination of complex vocal signals. Throughout evolution there has been a reciprocal relationship between hearing and acoustic communication, each system exerting pressure on the other and both together providing the basis for an effective social organization. The birds and primates are particularly good examples.

Study of the acoustic sense must proceed along many different routes. It is probably unnecessary to underscore the importance of physiological and anatomical mechanisms in accounting for the complex transformations that occur between changes in sound pressure at the outer ear and the final behavioral response of the animal to those changes. The biomechanics of the conducting pathway in the outer and middle ear, the transduction and coding of the stimulus in the inner ear, and the many influences along the ascending pathway and in the brain itself are the basic workings of the auditory system. They are all major contributors to the form and function of the final output, which is the behavior itself in response to acoustic stimulation.

This behavior, which we call hearing, has been the focus of these chapters. More broadly, the book has been concerned with the ways in which animals use sound in communicating with one another or in locating objects in their environment that emit or reflect sound. The contributing disciplines have been field and evolutionary biology and comparative sensory psychology or animal psychophysics. Both field and laboratory findings have been essential in putting together a preliminary account of comparative perception and communication. Unfortunately, different backgrounds and traditions have often prevented effective interactions between field and laboratory scientists. Laboratory investigations are apt to be considered too artificial, too analytic, and simply too far removed from the exigencies of the real world. Field studies, on the other hand, have been labeled descriptive and correlational rather than experimental, with too many conditions that vary concurrently and that therefore are not open to rigorous experimental questioning. As we have seen, both approaches have been successful and both have limitations. They must be combined in order to be most productive in the study of comparative perception, communication, and their biological relevance. Recently there have been fruitful collaborative efforts in which biological signals whose function has been studied in the field have been examined for their perceptual significance in the laboratory. Such productive interaction should be encouraged.

Recent theoretical and experimental developments in biology and experimental psychology raise issues that are germane to the study of hearing and acoustic communication in animals, and some of these deserve further examination. Results of studies in which psychologists have attempted to teach apes a human language in nonvocal form are under something of a cloud. Their future contribution seems uncertain. Some biologists have asked that we consider inquiring into the minds of animals and, as part of a legitimate scientific enterprise, examine in animals concepts such as self-awareness and consciousness. Of course, the burden rests heavily on the shoulders of these "cognitive ethologists" to formulate verifiable experimental questions. Perhaps such inquiry can help dispel what in the minds of many is a quantum jump from animal communication to human language. How extensively can we question other animals about their innermost thoughts? In fact, as I have shown, animal psychophysicists have for some time been asking animals to account for the sensations that they experience, and there is good reason to support the validity and reliability of their answers. The cognitive approach in psychology is gathering momentum and there is every reason to include animal communication and comparative perception within its purview, although this has not yet been done.

Kin selection and inclusive fitness have been rallying cries of the new field of sociobiology. If communication is adaptive, it is so because it benefits self and/or next of kin—the carriers of our genes. We cry "wolf" even if so doing puts us at considerable risk, because our close relatives can hear us and remove themselves from harm's way. Rather than considering the mutual benefits of communication to signaler and receiver, some sociobiologists stress the signaler as a manipulator of the receiver. Successful manipulation bestows a selective advantage on the signaler regardless of what happens to the receiver.

Studies of apes struggling to speak in voiceless modes, of animal thought and language, of cognitive psychology, and of the tenets of sociobiology may all have significant effects on our views and understanding of the acoustic sense of animals. It is quite clear that many different approaches are needed, for there continue to be sizable gaps in our knowledge of how animals hear and perceive and how they communicate with one another.

Field and laboratory studies of birds, bats, dolphins, and primates (human and nonhuman) have been particularly helpful in revealing how these animals make use of sound in detecting and discriminating among significant environmental events and in exchanging information with other animals. It is an auspicious beginning and brings with it the comprehension of how much more is still unknown. Where, for example, in animal communication can we find the origins of human speech and language? What are the many identifying characteristics of communication sounds that permit individual recognition, and yet enable animals to read the same message in vocal signals which vary significantly in many aspects of their acoustic structure? We know almost nothing about what reptiles hear or listen to or what they have to say to one another. Acoustic communication among fishes is more widespread than most of us ever suspected; only recently have researchers agreed that they could hear at all.

It would not be difficult to write a book or perhaps a catalog detailing all that awaits investigation. The opportunities for the interested scientist are considerable. This book was designed to provide readers with a summary of what is now known about the acoustic sense—but new discoveries are continually being made; the rate of progress in the field will be evident by what has changed since the book was completed. For the reader whose primary interests lie in related or distant fields, this book should provide some vital statistics as well as a feeling for the issues, concerns, and techniques of the scientists who claim as their bailiwick the comparative biology of hearing. For those specialists from other fields of acoustics, I have tried to define an area of inquiry and to assemble for the first time a reasonable sampling of its methods and accomplishments. This book is intended as an overview; if it has opened the door to further inquiry, it has been in some small measure successful.

SUGGESTED READINGS
ILLUSTRATION CREDITS
INDEX

SUGGESTED READINGS

GENERAL

Bullock, T. H., ed. 1977. *Recognition of Complex Acoustic Signals*. Berlin: Dahlem Konferenzen.

Dawkins, R., and J. R. Krebs. 1978. Animal signals: Information or manipulation? In *Behavioural Ecology: An Evolutionary Approach*, ed. J. R. Krebs and N. B. Davies. Sunderland, Mass.: Sinauer Associates. Pp. 282–309.

Lewis, D. B., and D. M. Gower. 1980. *Biology of Communication*. New York: John Wiley.

Marler, P., and W. J. Hamilton. 1966. *Mechanisms of Animal Behavior*. New York: John Wiley. Chapters 11 and 12 on hearing and acoustic communication.

Popper, A. N., and R. R. Fay, eds. 1980. *Comparative Studies of Hearing in Vertebrates*. New York: Springer-Verlag.

Sebeok, T. A., ed. 1977. *How Animals Communicate*. Bloomington: Indiana University Press.

Stebbins, W. C.; C. H. Brown; and M. R. Petersen. Sensory processes in animals. In *Handbook of Physiology: Sensory Processes*, vol. 1, ed. I. Darian-Smith, J. Brookhart, and V. B. Mountcastle. Washington, D.C.: American Physiological Society. In press.

Van Bergeijk, W. A.; J. R. Pierce; and E. E. David. 1960. *Waves and the Ear*. New York: Doubleday.

Yost, W. A., and D. W. Nielsen. 1977. *Fundamentals of Hearing*. New York: Holt, Rinehart, and Winston.

1. SOUND AND HEARING

Albers, V. M. 1970. *The World of Sound*. Cranbury, N.J.: A. S. Barnes.

Beranek, L. L. 1949. *Acoustic Measurements*. New York: John Wiley.

Keidel, W. D., and W. D. Neff, eds. 1974. *Handbook of Sensory Physiology: The Auditory System—Anatomy, Physiology (Ear)*, vol. 5, pt. 1. Berlin: Springer-Verlag. Chapter 2 on the acoustic stimulus.

Stevens, S. S., and F. Warshofsky. 1971. *Sound and Hearing.* New York: Time-Life Books.

Yost, W. A., and D. W. Nielsen. 1977. *Fundamentals of Hearing.* New York: Holt, Rinehart, and Winston. Chapters 1–3 on the physics of sound.

2. INSECTS: THE AUDIBLE INVERTEBRATES

Dethier, V. G. 1963. *The Physiology of Insect Senses.* London: Chapman and Hall. Chapter 4 on sound reception.

Haskell, P. T. 1961. *Insect Sounds.* London: Witherby.

Keidel, W. D., and W. D. Neff, eds. 1974. *Handbook of Sensory Physiology: The Auditory System—Anatomy, Physiology (Ear)*, vol. 5, pt. 1. Berlin: Springer-Verlag. Chapter 12 on invertebrate hearing.

Michelsen, A. 1973. The mechanics of the locust ear: An invertebrate frequency analyzer. In *Basic Mechanisms in Hearing,* ed. A. R. Møller. New York: Academic Press. Pp. 911–931.

Roeder, K. D. 1967. *Nerve Cells and Insect Behavior,* 2nd ed. Cambridge, Mass.: Harvard University Press.

3. THE ENIGMATIC FISHES

Hoar, W. S., and D. J. Randall, eds. 1971. *Fish Physiology: Sensory Systems and Electric Organs,* vol. 5. New York: Academic Press. Chapters 6–8 on sound production, hearing, and the morphology and physiology of inner ear and lateral line organs in fishes.

Hodgson, E. S., and R. F. Mathewson, eds. 1978. *Sensory Biology of Sharks, Skates and Rays.* Arlington, Va.: Office of Naval Research, Department of the Navy.

Keidel, W. D., and W. D. Neff, eds. 1974. *Handbook of Sensory Physiology: The Auditory System—Anatomy, Physiology (Ear)*, vol. 5, pt. 1. Berlin: Springer-Verlag. Chapters 3, 6, and 13 on middle and inner ear morphology and evolution.

Popper, A. N., and S. Coombs. 1980. Auditory mechanisms in teleost fishes. *American Scientist* 68:429–440.

Popper, A. N., and R. R. Fay, eds. 1980. *Comparative Studies of Hearing in Vertebrates.* New York: Springer-Verlag. Chapters 1–3 on fish hearing.

Tavolga, W. N.; A. N. Popper; and R. R. Fay, eds. 1981. *Hearing and Sound Communication in Fishes.* New York: Springer-Verlag.

4. AMPHIBIANS, REPTILES, AND BIRDS

Capranica, R. R. 1976. Morphology and physiology of the auditory system. In *Frog Neurobiology: A Handbook,* ed. R. Llinas and W. Precht. New York: Springer-Verlag. Pp. 551–575.

Garrick, L. D.; J. W. Lang; and H. A. Herzog. 1978. Social signals of adult American alligators. *Bulletin of the American Museum of Natural History* 160:157–192.

Keidel, W. D., and W. D. Neff, eds. 1974. *Handbook of Sensory Physiology: The Auditory System—Anatomy, Physiology (Ear)*, vol. 5, pt. 1. Berlin: Springer-Verlag. Chapters 3, 6, and 13 on middle and inner ear morphology and evolution.

Marcellini, D. L. 1978. The acoustic behavior of lizards. In *Behavior and Neurology of Lizards*, ed. N. Greenberg and P. D. MacLean. Rockville, Md.: Department of Health, Education, and Welfare, National Institute of Mental Health. Pp. 287–300.

Popper, A. N., and R. R. Fay, eds. 1980. *Comparative Studies of Hearing in Vertebrates*. New York: Springer-Verlag. Chapters 4–11 on hearing and the auditory systems of amphibians, reptiles, and birds.

Thielcke, G. A. 1976. *Bird Sounds*. Ann Arbor: University of Michigan Press.

Wever, E. G. 1978. *The Reptile Ear: Its Structure and Function*. Princeton: Princeton University Press.

5. THE TERRESTRIAL MAMMALS

Heffner, H., and B. Masterton. 1980. Hearing in glires: Domestic rabbit, cotton rat, feral house mouse, and kangaroo rat. *Journal of the Acoustical Society of America* 68:1584–99.

Keidel, W. D., and W. D. Neff, eds. 1974. *Handbook of Sensory Physiology: The Auditory System—Anatomy, Physiology (Ear)*, vol. 5, pt. 1. Berlin: Springer-Verlag. Chapters 3 and 5 on middle and inner ear morphology.

Peters, R. 1980. *Mammalian Communication: A Behavioral Analysis of Meaning*. Monterey: Brooks Cole.

Popper, A. N., and R. R. Fay, eds. 1980. *Comparative Studies of Hearing in Vertebrates*. New York: Springer-Verlag. Chapters 12 and 15 on hearing and the auditory systems of mammals.

Stebbins, W. C., ed. 1970. *Animal Psychophysics: The Design and Conduct of Sensory Experiments*. New York: Appleton-Century-Crofts.

Webster, D. B., and M. Webster. 1972. Kangaroo rat auditory thresholds before and after middle ear reduction. *Brain, Behavior, and Evolution* 5:41–53.

6. AERIAL AND AQUATIC MAMMALS

Busnel, R. G., and J. F. Fish, eds. 1980. *Animal Sonar Systems*. New York: Plenum Press.

Griffin, D. R. 1958. *Listening in the Dark*. New Haven: Yale University Press.

Herman, L. M., ed. 1980. *Cetacean Behavior*. New York: John Wiley. Chapters 1 and 4 on cetacean hearing and communication.

Kellogg, W. N. 1961. *Porpoises and Sonar*. Chicago: University of Chicago Press.

Long, G. R., and H. U. Schnitzler. 1975. Behavioural audiograms from the bat, *Rhinolophus ferrumequinum*. *Journal of Comparative Physiology* 100:211–219.

Peters, R. 1980. *Mammalian Communication: A Behavioral Analysis of Meaning*. Monterey: Brooks Cole. Chapter 7 on dolphin communication.

Sales, G. D., and J. D. Pye. 1974. *Ultrasonic Communication by Animals*. London: Chapman and Hall.

Simmons, J. A., and J. A. Vernon. 1971. Echolocation: Discrimination of targets by the bat, *Eptesicus fuscus. Journal of Experimental Zoology* 176:315–328.

7. THE DISCRIMINATING PRIMATES

Green, S. 1975. Variation of vocal pattern with social situation in the Japanese monkey (*Macaca fuscata*): A field study. In *Primate Behavior,* vol. 4, ed. L. A. Rosenblum. New York: Academic Press. Pp. 1–102.

Morse, P. A., ed. 1979. The perception of species-specific vocalizations. *Brain, Behavior, and Evolution* 16:321–463.

Peters, R. 1980. *Mammalian Communication: A Behavioral Analysis of Meaning.* Monterey: Brooks Cole. Chapter 6 on primate communication.

Snowdon, C. T.; C. H. Brown; and M. R. Petersen, eds. *Primate Communication.* New York: Cambridge University Press. In press.

Stebbins, W. C., ed. 1970. *Animal Psychophysics: The Design and Conduct of Sensory Experiments.* New York: Appleton-Century-Crofts. Chapters 3, 4, 13, 14, and 15 on hearing in monkeys.

Stebbins, W. C., and S. D. Iverson, eds. 1978. Hearing and acoustic communication in primates. In *Recent Advances in Primatology: Behaviour,* vol. 1, ed. D. J. Chivers and J. Herbert. London: Academic Press. Pp. 701–852.

Seyfarth, R. M.; D. L. Cheney; and P. Marler. 1980. Vervet monkey alarm calls: Semantic communication in a free-ranging primate. *Animal Behavior* 28:1070–94.

ILLUSTRATION CREDITS

1.2   W. Burns, *Noise and Man* (Philadelphia: Lippincott, 1968).

1.3   S. Green, Variation of vocal pattern with social situation in the Japanese monkey (*Macaca fuscata*): A field study, in L. A. Rosenblum, ed., *Primate Behavior: Developments in Field and Laboratory Research,* vol. 4 (New York: Academic Press, 1975).

2.1   H. Autrum, Schallempfang bei Tier und Mensch, *Naturwissenschaften* 30(1942):69–85.

2.2   H. Risler, Das Gehörorgan der Männchen von *Culex pipiens* L., *Aedes aegypti* L. und *Anopheles stephensi* Liston: eine vergleichend morphologische Untersüchung, *Zoologisches Jahrbuch, Abteilung Anatomie und Ontogenie der Tiere* 74(1955):478–490.

2.3   B. Burnet, K. Connolly, and L. Dennis, The function and processing of auditory information in the courtship behavior of *Drosophila melangaster, Animal Behavior* 19(1971):409–415.

2.4   H. Ghiradella, Fine structure of the noctuid moth ear, I: The transducer area and connections to the tympanic membrane in *Feltia subgothica* Haworth, *Journal of Morphology* 134(1971):21–45.

2.5   K. D. Roeder, *Nerve Cells and Insect Behavior,* 2nd ed. (Cambridge, Mass.: Harvard University Press, 1967).

2.6   R. S. Payne, K. D. Roeder, and J. Wallman, Directional sensitivity of the ears of noctuid moths, *Journal of Experimental Biology* 44(1966):17–31.

2.7   A. Michelsen, The physiology of the locust ear, II: Frequency discrimination based on resonances in the tympanum, *Zeitschrift für Vergleichende Physiologie* 71(1971):63–101.

2.8   M. Wells, *Lower Animals* (New York: McGraw-Hill World University Library, 1968); copyright © McGraw-Hill, used with the permission of McGraw-Hill Book Company.

3.2   G. Retzius, Das Gehörorgan der Wirbeltiere, I: *Das Gehörorgan der Fische und Amphibien* (Stockholm: Samson and Wallin, 1881).

3.3   E. G. Wever, The evolution of vertebrate hearing, in W. D. Keidel and W. D. Neff, eds., *Handbook of Sensory Physiology: The Auditory System—Anatomy, Physiology (Ear)*, vol. 5, pt. 1 (Berlin: Springer-Verlag, 1974).

3.5   K. von Frisch, Über den Gehörsinn der Fische, *Biological Reviews* 11 (1936):210–246.

3.6   S. Coombs and A. N. Popper, Hearing differences among Hawaiian squirrelfish (family Holocentridae) related to differences in the peripheral auditory system, *Journal of Comparative Physiology* 132 (1979):203–207. L. J. Sivian and S. D. White, On minimum audible fields, *Journal of the Acoustical Society of America* 4 (1933):288–321.

3.7   S. Coombs and A. N. Popper, Hearing differences among Hawaiian squirrelfish (family Holocentridae) related to differences in the peripheral auditory system, *Journal of Comparative Physiology* 132 (1979):203–207.

3.8   A. D. Hawkins and O. Sand, Directional hearing in the median vertical plane by the cod, *Journal of Comparative Physiology* 122 (1977):1–8.

4.1   E. G. Wever, The evolution of vertebrate hearing, in W. D. Keidel and W. D. Neff, eds., *Handbook of Sensory Physiology: The Auditory System—Anatomy, Physiology (Ear)*, vol. 5, pt. 1 (Berlin: Springer-Verlag, 1974).

4.2   C. L. Whitney and J. R. Krebs, Mate selection in Pacific tree frogs, *Nature* 255 (1975):325–326; reprinted by permission from *Nature,* copyright © 1975 Macmillan Journals Limited.

4.3   E. G. Wever, Structure and function of the lizard ear, *Journal of Auditory Research* 5 (1965):331–371.

4.4   I. L. Baird, Anatomical features of the inner ear in submammalian vertebrates, in W. D. Keidel and W. D. Neff, eds., *Handbook of Sensory Physiology: The Auditory System—Anatomy, Physiology (Ear)*, vol. 5, pt. 1 (Berlin: Springer-Verlag, 1974).

4.5   E. G. Wever, *The Reptile Ear: Its Structure and Function* (Princeton: Princeton University Press, 1978); copyright © 1978 by Princeton University Press, reprinted by permission of Princeton University Press.

4.6   W. C. Patterson, Hearing in the turtle, *Journal of Auditory Research* 6 (1966):453–464.

4.7   D. L. Marcellini, The acoustic behavior of lizards, in N. Greenberg and P. D. MacLean, eds., *Behavior and Neurology of Lizards* (Rockville, Md.: Department of Health, Education, and Welfare, National Institute of Mental Health, 1978).

4.9   R. J. Dooling, Behavior and psychophysics of hearing in birds, in A. N. Popper and R. R. Fay, eds., *Comparative Studies of Hearing in Vertebrates* (New York: Springer-Verlag, 1980).

4.11   P. Marler, Bird songs and mate selection, in W. E. Lanyon and W. N. Tavolga, eds., *Animal Sounds and Communication* (Washington, D.C.: American Institute of Biological Sciences, 1960); copyright 1960 by the American Institute of Biological Sciences.

4.12   P. Marler and M. Tamura, Culturally transmitted patterns of vocal behavior in sparrows, *Science* 146 (1964):1483–86; copyright 1964 by the American Association for the Advancement of Science.

4.13   G. A. Thielcke, *Bird Sounds* (Ann Arbor: University of Michigan Press, 1976).

5.2   Drawing courtesy of J. E. Hawkins, Jr.

5.3   H. Heffner and B. Masterton, Hearing in glires: Domestic rabbit, cotton rat, feral house mouse, and kangaroo rat, *Journal of the Acoustical Society of America* 68 (1980):1584–99. D. B. Webster and M. Webster, Kangaroo rat auditory thresholds before and after middle ear reduction, *Brain, Behavior, and Evolution* 5 (1972):41–53.

5.6   C. A. Prosen, M. R. Petersen, D. B. Moody, and W. C. Stebbins, Auditory thresholds and kanamycin-induced hearing loss in the guinea pig assessed by a positive reinforcement procedure, *Journal of the Acoustical Society of America* 63 (1978):559–566.

5.7   Photograph by Roger Peters, courtesy of the Chicago Zoological Society. Sonograph courtesy of Fred Harrington.

5.8   Sonograph courtesy of Preston Somers.

6.1   G. Sales and D. Pye, *Ultrasonic Communication by Animals* (London: Chapman and Hall, 1974).

6.2   G. Sales and D. Pye, *Ultrasonic Communication by Animals* (London: Chapman and Hall, 1974).

6.3   J. I. Dalland, Hearing sensitivity in bats, *Science* 150 (1965):1185–86; copyright 1965 by the American Association for the Advancement of Science.

6.4   G. R. Long and H. U. Schnitzler, Behavioral audiograms from the bat *Rhinolophus ferrumequinum, Journal of Comparative Physiology* 100 (1975):211–219.

6.5   C. S. Johnson, Auditory thresholds of the bottlenose dolphin, *Tursiops truncatus* Montagu (China Lake, Calif.: U.S. Naval Ordinance Test Station NOTSTP 4178, 1966).

6.6   W. W. L. Au, Echolocation signals of the Atlantic bottlenose dolphin (*Tursiops truncatus*) in open waters, in R-G. Busnel and J. F. Fish, eds., *Animal Sonar Systems* (New York: Plenum Press, 1980).

6.7   J. A. Simmons and J. A. Vernon, Echolocation: Discrimination of targets by the bat *Eptesicus fuscus, Journal of Experimental Zoology* 176 (1971):315–328.

6.8   J. A. Simmons and J. A. Vernon, Echolocation: Discrimination of targets by the bat *Eptesicus fuscus, Journal of Experimental Zoology* 176(1971):315–328.

6.9   A. E. Murchison, Detection range and range resolution of echolocating bottlenose porpoise (*Tursiops truncatus*), in R-G. Busnel and J. F. Fish, eds., *Animal Sonar Systems* (New York: Plenum Press, 1980).

7.1   J. H. Elder, Auditory acuity of the chimpanzee, *Journal of Comparative Psychology* 17(1934):157–183; copyright 1934 by The Williams & Wilkins Co., Baltimore. D. N. Farrer and M. M. Prim, A preliminary report on auditory frequency threshold comparisons of humans and pre-adolescent chimpanzees, Technical Report 65-6 (Holloman Air Force Base, N. Mex.: 6571st Aeromedical Research Laboratory, 1965). H. Heffner, R. Ravizza, and B. Masterton, Hearing in primitive mammals, III: Tree shrew (*Tupaia glis*), *Journal of Auditory Research* 9(1969a):12–18. H. Heffner, R. Ravizza, and B. Masterton, Hearing in primitive mammals, IV: Bushbaby (*Galago senegalensis*), *Journal of Auditory Research* 9(1969b):19–23. L. J. Sivian and S. D. White, On minimum audible fields, *Journal of the Acoustical Society of America* 4(1933):288–321. W. C. Stebbins, S. Green, and F. L. Miller, Auditory sensitivity of the monkey, *Science* 153 (1966):1646–47; copyright 1966 by the American Association for the Advancement of Science. W. C. Stebbins, Hearing, in A. M. Schrier and F. Stollnitz, eds., *Behavior of Nonhuman Primates,* vol. 3 (New York: Academic Press, 1971).

7.2   W. C. Stebbins, Hearing of Old World monkeys (Cercopithecinae), *American Journal of Physical Anthropology* 38(1973):357–364.

7.3   J. O. Nordmark, Mechanisms of frequency discrimination, *Journal of the Acoustical Society of America* 44(1968):1533–40. W. C. Stebbins, Hearing of Old World monkeys (Cercopithecinae), *American Journal of Physical Anthropology* 38(1973):357–364.

7.4   S. Filling, *Difference Limen for Frequency: Studies on a Series of Normal Subjects and on a Series of Patients from a Hearing Rehabilitation Centre* (Odense, Denmark: Andelsbogtrykkeriet, 1958). H. Heffner, R. Ravizza, and B. Masterton, Hearing in primitive mammals, III: Tree shrew (*Tupaia glis*), *Journal of Auditory Research* 9(1969a):12–18. H. Heffner, R. Ravizza, and B. Masterton, Hearing in primitive mammals, IV: Bushbaby (*Galago senegalensis*), *Journal of Auditory Research* 9(1969b):19–23. W. C. Stebbins, Hearing of the primates, in D. Chivers and J. Herbert, eds., *Recent Advances in Primatology,* vol. 1, *Primate Behavior* (London: Academic Press, 1978); copyright by Academic Press Inc. (London) Ltd.

7.6   C. H. Brown, M. D. Beecher, D. B. Moody, and W. C. Stebbins, Localization of noise bands by Old World monkeys, *Journal of the Acoustical Society of America* 68(1980):127–132.

7.8   C. H. Brown, M. D. Beecher, D. B. Moody, and W. C. Stebbins, Localization of primate calls by Old World monkeys, *Science* 201(1978):753–754; copyright 1978 by the American Association for the Advancement of Science.

7.9   S. Green, Variation of vocal pattern with social situation in the Japanese monkey (*Macaca fuscata*): A field study, in L. A. Rosenblum, ed., *Primate Behavior: Developments in Field and Laboratory Research,* vol. 4, (New York: Academic Press, 1975).

Index